Springer Series in Synergetics

Editor: Hermann Haken

Synergetics, an interdisciplinary field of research, is concerned with the cooperation of individual parts of a system that produces macroscopic spatial, temporal or functional structures. It deals with deterministic as well as stochastic processes.

Volumes 1–37 are listed at the end of the book

Yurii A. Kravtsov (Ed.)

Limits of Predictability

With 62 Figures

Springer-Verlag Berlin Heidelberg GmbH

Professor Dr. Yurii A. Kravtsov

Institute of General Physics, Small Venture GROT, Russian Academy of Sciences,
Vavilov Street 38, Moscow 117942, Russia

Series Editor:

Professor Dr. Dr. h. c. Hermann Haken

Institut für Theoretische Physik und Synergetik der Universität Stuttgart,
D-70550 Stuttgart, Germany and
Center for Complex Systems, Florida Atlantic University,
Boca Raton, FL 33431, USA

ISBN 978-3-642-51010-6   ISBN 978-3-642-51008-3 (eBook)
DOI 10.1007/ 978-3-642-51008-3

Library of Congress Cataloging-in-Publication Data. Limits of predictability / [edited by] Yurii A. Kravtsov. p. cm.
– (Springer series in synergetics; v. 60) Includes bibliographical references and index.
1. Prediction theory. I. Kravtsov, Yurii
Aleksandrovich. II. Series. QA279.2.L56   1993   003'.2–dc20   93-7284

© Springer-Verlag Berlin Heidelberg 1993
Softcover reprint of the hardcover 1st edition 1993

The use of general descriptive names, registered names, trademarks, etc. in this publication does not imply, even in the absence of a specific statement, that such names are exempt from the relevant protective laws and regulations and therefore free for general use.

Typesetting and Layout: J. Andrew Ross
57/3140 - 5 4 3 2 1 0 - Printed on acid-free paper

Preface

This book addresses the problem of predictability of various phenomena, both of physical origin (such as weather, climate, earthquakes, biological media, and dynamical chaos) and of a social nature (election preferences, laws of ethnogenesis, and so on).

The book explores the predictive power of modern science, and consists of a set of survey chapters by distinguished experts, who have written them in a style that is understandable to nonexperts. The importance of the problems under review and the academic distinction of the team of authors ensure that this book will attract a large audience of readers who are interested in learning more about the achievements and prospects of modern science.

All the authors have sought to make their articles popular enough to be understood by the average mathematically literate reader. At the same time, however, the articles discuss a number of fundamental questions that deserve expert attention. I hope we have managed to meet both these requirements, which are so hard to reconcile.

I am very grateful to all the contributors, who eagerly accepted the request to write chapters for this book and completed the work within a very tight deadline. I would like to thank M. G. Makhova and A. V. Karaseva for assistance in readying the manuscript for publication, to S. N. Gonshorek for efforts that helped this book see the light, to E. B. Grigoreva and A. A. Starkov for their timely and expert translation of the book, and to J. A. Ross at Springer-Verlag for copy-editing the text and typing the camera-ready copy.

Moscow, January 1993   Yu. A. Kravtsov

Contents

Contributors

I. V. Bestuzhev-Lada
    Institute of Sociology
    Russian Academy of Sciences
    Moscow, Russia

L. N. Gumilev
    Institute of Geography
    St. Petersburg, Russia

G. R. Ivanitskii
    Institute of Biophysics
    Russian Academy of Sciences
    Pushchino
    Moscow District, Russia

V. I. Keilis-Borok
    Institute for Earthquake Prediction Theory
    and Mathematical Geophysics
    Russian Academy of Sciences
    Moscow, Russia

Yu. A. Kravtsov
    Institute of General Physics, Small Venture GROT
    Russian Academy of Sciences
    Vavilov Street 38
    Moscow 117942, Russia

A. J. Lichtman
    Department of History
    The American University
    Washington, D.C.
    United States of America

V. A. Lisichkin
PROGNOSTIKA Science and Production Center
Russian Academy of Sciences
Moscow, Russia

G. G. Malinetskii
Institute of Applied Mathematics
Russian Academy of Sciences
Moscow, Russia

A. S. Monin
Institute of Oceanology
Russian Academy of Sciences
Moscow, Russia

V. F. Pisarenko
Institute for Earthquake Prediction Theory
and Mathematical Geophysics
Russian Academy of Sciences
Moscow, Russia

L. I. Piterbarg
Institute of Oceanology
Russian Academy of Sciences
Moscow, Russia

M. A. Sadovskii
Institute of Earth Physics
Russian Academy of Sciences
Moscow, Russia

Yu. Yermolaev
Northwestern Division
Central Economic Institute of Russia
St. Petersburg, Russia

... Destiny is exactly what I don't believe in. I think the future is unpredictable and vague, and that it is framed by all of us, bit by bit through our infinitely complex interactions.

A. D. Sakharov

1. Introduction

Yu. A. Kravtsov

This book examines the limits of predictability in various fields of knowledge. A matter of interest in all times, the question of predictability has acquired particular importance now that the countries of the former Soviet Union and Eastern Europe have entered an era of dramatic political and economic reforms.

A mere 30–50 years ago it was maintained that scientific forecasting had limitless opportunities. This idea was rooted in the 'romantic' era of science (the nineteenth century and the early twentieth century) when astounding progress in science and engineering did not let one see that any prediction is limited in principle.

The situation today is very different. It has become clear to many – the brightest experts were aware of this long ago – that long-term forecasts are virtually impossible to make. This applies also to the exact sciences, including physics, the 'classical' part of which once firmly supported Laplace's determinism. Physicists' loyalty to the possibility of forecasting for macroscopic systems has been completely shattered with the discovery of dynamical chaos and other, similar phenomena. In dynamical systems showing chaotic behavior, microscopic disturbances rather quickly acquire macroscopic values (by an exponential law) and thereby make it impossible to predict the behavior of many dynamic systems even for short periods of time. As a result, research into chaotic systems has led to a fundamental revision of the entire classical concept of predictability.

Indeed, who could have thought a mere 30–40 years ago that a slight disturbance of the atmosphere in South America (the stroke of a butterfly's wings) can bring about a change in the weather over Siberia a month or two later? The high degree of interrelatedness of events in the world of physics – 'everything affects everything' – prompts one to reconsider the customary ideas of the requirements to be met to make an intricate physical system behave in a predictable way. All such requirements prove to be extremely strict; as a result, few systems in the world around us can be called predictable.

The limits of a prognosis in time are aptly described by the term 'predictability horizon' coined recently by Sir James Lighthill. It is beyond any doubt that there must be predictability horizons not only in physics but also in all other fields of knowledge. One of the main aims of this book is to highlight the factors limiting the reliability of forecasting in various fields of human activity.

This is a difficult problem, for the models used in forecasting are as subjective as researchers' subjective ideas about this or that phenomenon. For the same reason, the terms 'scientific' and 'objective' that are used rather often to describe prognoses are in most cases no more than expressions of a serious, professional attitude to the work and the use of all available knowledge.

Social sciences, more than others, need convincing yardsticks to measure the authenticity of prognoses. For a long time in the Soviet Union, the social sciences relied on ideas that had very little to do with real science and were indeed utopian. It would probably make no sense to waste time and effort criticizing the methodological basis of Marxist prognoses that have not materialized. It is far more important to formulate the criteria that can help increase the accuracy of prognoses. In sociology this will be very hard to do, because any prognosis can itself affect, directly or indirectly, the measures taken either to implement it or to prevent it from happening. This is the fundamental difference between forecasting the behavior of 'unthinking' objects and forecasting for social entities, capable of conscious response to information.

This book reviews current approaches to evaluating the limits of predictability, and describes some principles of predictability and the models of prognosis used most frequently. Special attention is paid to the causes that affect the quality of prognoses: noise and disturbances, imperfect theoretical models (errors in models can be termed 'ignorance noise'), unavoidable uncertainty, misinformation, and so on. The factors which most interfere with prognostication at the current level of knowledge play a significant role in the book.

Scientists around the globe have already published a number of collections of works on this subject. Especially worthy of note are the book *Long-Term Predictability in Dynamics* edited by Shebenely and Taply [1.1] and the discussion on predictability published in the Proceedings of the Royal Society [1.2] in which both natural scientists (physicists, chemists, meteorologists, and biophysicists) and scholars (economists and philosophers) took part.

This is an interdisciplinary book and consists of nine chapters on various fields of knowledge. Six chapters are devoted to the problem of predictability in natural sciences.

Chapter 2 is a fundamental article by A. S. Monin and L. I. Piterbarg on forecasting weather and climate. The article gives an insight into how weather and climatic forecasts are made and highlights the factors that hinder long-term forecasts. The main such factor is the local instability of air currents, which imposes fundamental restrictions on the durability and quality of prognoses. This is also true of many other physical phenomena.

Chapter 3, by G. R. Ivanitskii, is written in a different key. The author gives an articulate description of prediction-related properties of an active biological medium. Ivanitskii believes that these properties imitate the ability of the brain (which is also an active biological medium) to predict the future even if the signal-to-noise ratio is not very favorable.

In Chapter 4, G. G. Malinetskii interprets the problem of predictability from the standpoint of synergetics – the science of self-organizing behavior in complex systems. In a style that is easy but not trivializing, Malinetskii leads the reader into the world of nonlinear phenomena, which displays an unusual variety of modes of behavior and provides the methodological basis for prognoses concerning complicated dynamical systems.

Chapter 5, by V. A. Lisichkin, is devoted to the theoretical and informational interpretation of forecasting. Lisichkin considers a prognosis as a signal, sent from the future into the present, which is vulnerable (like any signal) to certain kinds of noise and interference. The information-theoretic approach paves the way for a quantitative description of the authenticity of a prognosis.

In Chapter 6, M. A. Sadovskii and V. F. Pisarenko analyze the sources of accidentality in forecasting and offer several forecasting patterns based on the analyses of time series. Although the quoted examples primarily concern the predictability of earthquakes (the authors are leading experts in that field), the findings are widely applicable.

My own contribution, Chapter 7, is devoted to the factors limiting the horizon of predictability. It turns out that the horizon of predictability is very rarely limited by fundamental factors – quantum fluctuations. In most cases the restrictions are the result of a bad model of various microscopic fluctuations. The latter is particularly true of systems that are unstable locally. My work also considers some physical–social analogies (for instance, the increase of fluctuations during phase transitions) and emphasizes the limited applicability of these analogies to social forecasting.

The last three chapters of the book are devoted to social phenomena proper. In Chapter 8, I. V. Bestuzhev-Lada considers in detail the 'Oedipus effect,' in other words, the ability of an individual human being (or a social group) to resist or promote the materialization of predictions. This effect prompts one to take a totally different attitude to social forecasting than to other types of prognosis.

In Chapter 9, V. I. Keilis-Borok and A. J. Lichtman review the results of a successful attempt to detect factors that have a significant impact on a specific social phenomenon – returns from presidential and senatorial elections in the U.S.A. They find that those factors which have for a long time been considered crucial (such as a candidate's behavior during the election campaign) are in fact of secondary importance, whereas those considered secondary are in fact crucial. The authors are scholars representing two leading forecasting organizations in Russia and the U.S.A. Their work shows the universality of some forecasting techniques (such as correlation analysis of indicators, which works successfully in the social sphere, geophysics, and earthquake forecasting) and is a good example of international cooperation in science.

The book ends with Chapter 10, an article by an outstanding fellow Russian, the founder of ethnology L. N. Gumilev, who together with his disciple and follower V. Yu. Yermolaev has made an interesting attempt to highlight the similarities between ethnogenesis and a self-organizing system (a laser). Although the analogy is found only at the qualitative level so far, and there has been no confirmation at the level adopted in the natural sciences, the attempt itself should be regarded as a success, because it prompts dramatically new approaches to forecasting the future of ethnic divisions. It is certain that this new work by Gumilev and

Yermolaev will be of interest not only to people trying to overcome, together or on their own, the legacy of totalitarian rule, and who have already entered the critical phase of realizing their place in the history of humanity, but also to other people who are aware of their responsibility for the future of civilization.

References

1.1 V. Shebenely, B.D. Taply (eds.): Long-Term Predictability in Dynamics. Dordrecht: Reidel, 1976
1.2 Discussion on principles of predictability. Proceedings of the Royal Society A407(1832), 1–238 (1986)

2. Forecasting Weather and Climate

A. S. Monin and L. I. Piterbarg

2.1 Weather and Climate

The reader is already well acquainted with the subject of this chapter through the daily reports of his or her national weather forecasting service. Also, newspapers publish week-ahead weather forecasts that quite frequently come true. One may perhaps wonder why they do not forecast a month or two in advance. This would undoubtedly be of great importance to many fields of activity, especially those related to agriculture. With a reliable long-term weather forecast at hand, sowing, fertilizing, and harvesting could be properly scheduled.

Besides, wouldn't it be good to know what kind of weather we shall experience on the Earth in, say, a hundred or a thousand years? At this point, of course, it is more appropriate to speak of weather averaged over a long period of time, namely climate (a definition is given below). Climate forecasting is also one of the critical problems for fundamental science. So why do our forecasting services not provide us with monthly, yearly, and centennial forecasts? Is it because they are short of modern computers or are their experts not skilled enough?

Alas, the reasons are much more deeply rooted. When it is considered seriously, the question of reliable forecasting inevitably leads to fundamental problems concerning the organization of nature that are tackled nowadays by theoretical physicists. To realize the complexity of the question, one needs to develop an insight into modern mathematical methods of predicting dynamical and stochastic systems. One of the aims of this chapter is to show that our everyday worries (like 'will it rain tomorrow?') and the problems exercising some of the greatest minds in science are not so widely separated.

When predicting the behavior of some physical quantity x (in the case of weather forecasting, it may be air temperature, wind velocity, amount of precipitation, etc.) we presume that this quantity is time dependent: $x = x(t)$. The forecast objective is to compute, predict, or guess quantity x for some future instant $t + \tau$ based on the behavioral history of both x and other quantities that determine the evolution of x and have influenced it up to time instant t, which can be associated with the present. We shall take τ to be the term of the forecast.

Such a definition of the problem implies that the evolution of $x(t)$ conforms to certain laws. These laws may be expressed both in deterministic form, through either differential or more complex integro-differential, operator-type, evolution equations, and in statistical form with the time-invariant or the statistical (following known laws) characteristics of $x(t)$ preset. As has been noted by Tatarskii [2.1], one and the same process can be regarded as both deterministic and random depending on the problem to be solved and available information on the development of the process.

For example, a forecast for 1–3 days ahead of air temperature for a given site (a given observing station) is made by integrating equations of motion and transport. For such a forecasting term, atmospheric processes can be taken as adiabatic, i.e., one can neglect external energy sources and energy dissipation due to viscosity. These equations faithfully represent the adiabatic dynamics and there is no need for any alternative expression.

However, the deterministic approach is no longer consistent when tackling problems of longer-term forecasting of temperature averaged over large territories and long time periods. The problem is that averaged hydrodynamic quantities cannot be governed by exact, closed equations. Insofar as hydrodynamic equations are nonlinear, the averaged equations of transport incorporate turbulent heat fluxes, while the equations of motion incorporate turbulent momentum fluxes (Reynolds stresses). Closing averaged equations is one of the most intricate problems in turbulence theory and generally has no satisfactory solution [2.2]. Existing semi-empirical methods of parameterization of heat, humidity, and momentum fluxes are far from perfect and almost useless for solving long-term forecasting problems. Why turbulent heat and momentum fluxes cannot be predetermined long in advance will be explained below.

At the same time, lengthy temperature observations conducted at many stations for more than a century furnish enough information to analyze statistical (i.e., averaged) characteristics of the quantity under study and to produce a statistical model of the long-term behavior of air temperature at a given site or over a given region or over greater areas (e.g., mean temperature over the Northern Hemisphere). In this case, a purely statistical approach may be preferred to the deterministic one. The simple idea of considering atmospheric and oceanic processes as deterministic events for short terms (weather) and as random events for long terms (climate) is central to what is presented below.

To begin a more exact treatment, we think it appropriate to outline some of the essential ideas of the theory of dynamical systems and to consider factors impeding predictability of these systems, namely, instability and stochastization (Sect. 2.2). This information will help us to understand why there is a limit to the predictability of real atmospheric processes within the context of hydrodynamic models (Sect. 2.3). Section 2.4 comprises elements of the mathematical theory of random process prediction, in a linear frame of reference, and is a necessary preamble to Sect. 2.5 dealing with climate predictability. Section 2.6 addresses the question of improving statistical forecasting and defines some limits to its efficiency (even in linear cases). Section 2.7 is concerned with the problem of applying weather forecasting results in the economy.

Let us assume that $\tau_0 = 1$ month is the temporal boundary distinguishing weather from climate. All events that take place within time scale τ_n much smaller than $\tau_0, \tau_n << \tau_0$, fall into the weather scale, while those occurring within time scale $\tau_k, \tau_k >> \tau_0$, refer to climate. We can regard a 'typical variability scale' as the correlation time. From this it follows that by averaging weather processes over a long period of time (a month and more) we obtain a climatic process. For instance, hour-to-hour variations of air temperature and humidity as well as of wind velocity feature a typical correlation time of 1–3 days, which means that such time series represent weather processes. Another example is sea surface temperature averaged over an upper mixed layer several tens of meters thick. Its evolution in time is obviously a climatic process.

2.2 Dynamical Systems and Their Properties

The most general model of natural processes is provided by a dynamical system. Suppose that the state of an object under investigation is prescribed at each instant of time by numerical values of parameters $p_1, p_2, \ldots$, whose set is written $\mathbf{p} = (p_1, p_2, \ldots)$ and called a state. The set of all possible (admissible) states $\mathbf{p} = \{p_i\}$ constitutes the phase space. If the variation of a system's state at successive instants can be computed on the basis of the evolution equation

$$\dot{\mathbf{p}} = \Phi(t,\mathbf{p}), \qquad\qquad (2.1)$$

where the dot denotes a time derivative and $\Phi(t,\mathbf{p})$ is a function in the phase space, then the system concerned is a dynamical system given by equation (2.1).

In case we are interested in studying only discrete-time states, the system evolution can be written as

$$\mathbf{p}(n+1) = \Phi_n[\mathbf{p}(n)],$$

where $\mathbf{p}(n)$ is the system's state at instant t_n, $n = 0,1,2,\ldots$

In studying natural processes, it is not usually possible to confine oneself to a finite (or even countable) set of parameters $p_1, p_2, \ldots$ defining the state. More often, one has a phase space whose elements are functions (or sets of functions) of time and space coordinates. Such systems are called continuous, or infinite-dimensional systems, in contrast to finite-dimensional systems whose states are given by finite parameter sets. For example, to give a mathematical description of a fluid flow state one uses functions governing fluid velocity distribution, $v(\mathbf{r},t)$, where $\mathbf{r}$ is the space coordinate, and two of the fluid's thermodynamic characteristics, e.g., pressure $p(\mathbf{r},t)$ and density $\rho(\mathbf{r},t)$. Equations giving the evolution of this field (a combination of Navier–Stokes and mass conservation equations) are so complicated that generally there are no proven uniqueness and existence theorems for them. In solving hydrodynamic, oceanographic, and meteorological problems, it often suffices to examine only the first terms of the field expansion in a series based on a complete set of functions. These functions are numbered in order of increasing analytic structure. Those coming first are termed lower-order modes, and the reduction of a continuous system to a finite-dimensional format, covering only the lower-order modes, is referred to as the Galerkin method.

Dynamical systems can be divided into autonomous (with Φ independent of t in the continuous case and of n in the discrete case) and non-autonomous (the contrary condition) systems. They can also be classified into linear systems (if Φ is a linear function of $\mathbf{p}$ in the finite-dimensional case or a linear operator over a functional space in the continuous case) and nonlinear ones (the contrary condition).

We shall restrict our discussion to autonomous systems, since models used in mechanics, hydrodynamics, and geophysics are commonly autonomous. Apart from this, one can always reduce a non-autonomous system to an autonomous one by expanding its phase space. Incidentally, this division in terms of linearity merits closer attention. We must note that linearity is not a very fortunate term here, for in fact there are no strictly linear systems in nature. We can regard a linear system as a limiting case of a nonlinear one if we assume that certain parameters are very small, or that some of the processes occurring can be neglected. For example, a linear pendulum equation is derived under the assumption that the deviation of the pendulum from its equilibrium position is small. Similarly, by assuming that the Reynolds number is small, i.e., neglecting inertia as compared with viscosity, we obtain linear equations for viscous fluids. The list of such examples can be extended. But why should one pursue this policy of substituting linear equations for nonlinear ones? There is only one reason: the analytical methods for investigating linear systems are more convenient. Problems in linear dynamics are indeed much easier to tackle than 'nonlinear' problems.

A clear physical distinction is made between linear and nonlinear systems. Let us consider the equation $\dot{p} = \Phi(p)$, where $p = p(\mathbf{r}, t)$ is a scalar function representing a state of a continuous system (e.g., let p be pressure at a point in space) and Φ is a linear operator with constant coefficients. Let $\hat{p}$ represent the Fourier transform:

$$p(\mathbf{r}, t) = \int \hat{p}(\mathbf{k}, t) \exp[i(\mathbf{k}, \mathbf{r})] d\mathbf{k}. \qquad (2.2)$$

Here $\mathbf{k}$ is a wave vector, i.e., we can imagine a time-varying field $p(\mathbf{r}, t)$ as a superposition of elementary sinusoidal waves with amplitudes $\hat{p}(\mathbf{k}, t)$. Then dynamic equations in the Fourier space (the space of wave vectors $\mathbf{k}$) assume the following simple form:

$$\frac{\partial}{\partial t} \hat{p}(\mathbf{k}, t) = \hat{\Phi}(\mathbf{k}) \, \hat{p}(\mathbf{k}, t), \qquad (2.3)$$

where $\Phi(\mathbf{k})$ is the image of operator Φ in the Fourier space. Thus we have obtained a continuous set of equations that are in no way connected and are each easily solved individually. In other words, waves superposed on one another to form the field $p(\mathbf{r}, t)$ do not interact at all; each evolves independently, its behavior determined only by its vector $\mathbf{k}$. Certainly, expression (2.2) holds for a field governed by nonlinear equations, but in that case the amplitude equations are interconnected (which is of prime importance) and the elementary sinusoidal components interact with each other. This interaction results in the much more complex asymptotic behavior of nonlinear systems as compared to linear ones. As shown by equation (2.3), for example, a stationary state of the linear system $\hat{p}(\mathbf{k}) = \lim \hat{p}(\mathbf{k}, t)$, if it exists, cannot produce a 'continuous' spectrum in a wave number space, such a spectrum differs from zero only for a set of $\mathbf{k}$ such that $\Phi(\mathbf{k}) = 0$, i.e., the spectrum dimension is sure to be less than two for all typical operators Φ. In the 'linear' case, the choice of possible behavior patterns for approximately stationary states (or periodic ones) is similarly small. At the same time, there is quite a variety of scenarios for the asymptotic behavior of nonlinear systems. Although these scenarios are very complicated, they follow certain general laws, which we shall now discuss.

The 'unusual' behavior of nonlinear dynamical systems was originally discovered by the prominent American theoretical meteorologist Edward Lorenz as he contemplated a problem concerned with weather forecasting. We shall briefly outline the history of his discovery, following [2.3]. The numerical charts for hydrodynamic short-term (several days ahead) weather forecasts that appeared in the mid-1950s proved to be inefficient, which led many investigators to turn their attention to statistical methods of forecasting based on the concept of linear regression (the methods are discussed below in greater detail). This development was in part spurred by the work of Wiener (see, e.g., [2.4]) on the problem of predicting stationary random processes.

It seemed as though application of a large number of predictors might supplant hydrodynamic forecasting techniques despite the fact that atmospheric processes display considerable nonlinearity. Lorenz was sceptical about the idea of statistical forecasting and decided to test its validity experimentally against a dynamical model. After laborious searching for a model to fit a chosen sample of aperiodic motion (it is clear that periodic progressions or ones like them are rather easy to predict without using sophisticated mathematics but merely using data about their history) Lorenz settled on a two-level model of the atmosphere. Using the Galerkin method of retaining only the most significant modes, the two-level model was reduced to a common system of differential equations. For a 12th-order system thus obtained, the linear statistical forecasting method was actually shown to be invalid.

In the course of performing this demonstration, however, Lorenz made a far more important discovery. While calculating one of the numerical solutions of the system, he printed out intermediate values of the phase variables that were truncated to three decimal places – whereas their representation stored in memory was continued to six decimal places. Taking these truncated values as the starting values for further computations and preparing a two-month-ahead forecast, Lorenz obtained results that were strikingly different from those obtained by integrating the system without using the truncated intermediate values and thereby losing the last three decimal places. At first he even suspected computer failure, but a careful check convinced him that rounding off initial data, innocuous as it seemed on the surface, actually affected the end results of the computation dramatically. Further analysis disclosed that initial noise of the order of 10^{-3} doubled every four model days or so and thus increased in two months by a factor of $2^{15} \approx 3.3 \cdot 10^4$, becoming several tens of units. After a while the model he used was simplified considerably to leave only three independent variables, and the famous Lorenz system was formulated:

$$\dot{X} = -\sigma X + \sigma Y,$$
$$\dot{Y} = rX - Y - XZ, \qquad (2.4)$$
$$\dot{Z} = bZ + XY.$$

The meaning of the variables is as follows. X characterizes the intensity of convective motions, Y is the temperature difference between upward and downward convective currents, and Z gives the deviation of the vertical temperature profile from a linear one. The fixed parameters are r, a relative Rayleigh number, σ, the Prandtl number, and b, a number expressing the vertical section geometry of convective billows.

Lorenz did not merely state the instability of solutions to system (2.4). Stability is understood here in the sense of Lyapunov: the phase trajectory $\mathbf{p} = \mathbf{p}(\mathbf{p}_0, t)$ (i.e., the solution to equation (2.1)) conforming to initial value $\mathbf{p}_0$ is stable if for any positive number ε, as small as desired, a corresponding positive number δ exists such that for any trajectory $\tilde{\mathbf{p}} = \tilde{\mathbf{p}}(\tilde{\mathbf{p}}_0, t)$ with $\tilde{\mathbf{p}}_0$ initially obeying the condition $\rho(\mathbf{p}_0, \tilde{\mathbf{p}}_0) < \delta$, where ρ is the distance between points $\mathbf{p}_0$ and $\tilde{\mathbf{p}}_0$, the inequality $\rho(\mathbf{p}, \tilde{\mathbf{p}}) < \varepsilon$ holds at any instant of time; that is, close initial conditions give rise to close phase trajectories. A more detailed analysis of solutions to system (2.4) led to the discovery of a subset of the phase space called a strange attractor where phase trajectories behave in a most peculiar manner.

To show how this works, we need to introduce several definitions that are less than perfectly exact from the mathematical point of view. Call a phase space point P

a *nonwandering* point if a phase trajectory setting out from any site in its neighborhood U returns to this site at least once. The nonwandering points are, for example, stationary points to which a trajectory adheres at any t values or points belonging to periodic orbits. If a phase trajectory $p(p_0, t)$ leaving some point p_0 within a phase space subset $N \subset P$ returns to it at any time $t > 0$, i.e., $p(p_0, t) \in N$, then such a set is said to be *invariant*. Invariant sets are exemplified by a stationary (limit) point and a periodic orbit. If N has no subsets with the said feature in common, then N is a *minimal invariant* set. Minimal invariant sets composed of nonwandering points are *attractors*. Finally, a *strange* attractor is an attractor that is neither a stationary point nor a periodic orbit. As the definition suggests, a strange attractor is 'home' to the aperiodic trajectories of a dynamical system.

A characteristic property of strange attractors occurring in systems of the hydrodynamic type (including the Lorenz system) is that the phase trajectories found in them display stochastic features, namely:

– An extremely sensitive dependence upon initial conditions (local instability) which is due to exponential divergence of initially close trajectories (which ultimately causes their unpredictability or unreproducibility from preset initial conditions given to any desired – but finite – accuracy);

– Denseness of trajectories throughout the attractor, implying that trajectories pass by any point in the attractor arbitrarily closely (and subsequently make close returns infinitely many times), with any initial non-equilibrium probability distribution (a measure) in the phase space (within the 'strange' attraction region, to be more exact) approaching a given limiting equilibrium distribution $P(A)$ on the attractor (an invariant measure);

– The feature of mixing, implying that for any (measurable) attractor's subsets A and B, the probability that a trajectory leaving A arrives at B in a long time is proportional to measure B:

$$\lim_{t \to \infty} P\{(F^t A) \cap B\} = P(A)P(B),$$

where $F^t A$ is a set of phase space points to which subset A of points arrive in time t in accordance with dynamic laws. Mixing implies independence of the time average $\langle F[p(t)] \rangle$ of any function $F(p)$ on the strange attractor from initial conditions p_0 (for almost any p_0) and its coincidence with the time invariant measure average value $\overline{F(p)}$ (ergodicity).

Mixing shows up as a rather fast decay of correlation functions over large times $(t \to \infty)$:

$$B_{ij}(\tau) = \left\langle \left[p_i(t) - \langle p_i(t) \rangle \right] \left[p_j(t + \tau) - \langle p_j(t + \tau) \rangle \right] \right\rangle$$

and hence continuity of their Fourier transform with respect to τ (spectra).

Systems investigated in hydrodynamics (including the Lorenz system) are dissipative, i.e., phase volume compression takes place within them. An exponential divergence of closely spaced trajectories during phase volume compression is possible if expansion taking place along some directions p_k of phase space P is compensated by compression taking place along other directions i.e., if nonwandering points are similar to two-dimensional saddle points. Such points are called hyperbolic, and in their neighborhood one finds a union of stable manifolds, where phase trajectories converge, and unstable manifolds, where they diverge exponentially. To what extent are these hyperbolic sets indicative of phase spaces in dynamical systems? Will a slight perturbation of such a system cause the hyperbolic sets to vanish? At this point, it is expedient to bring in the notion of a rough system (now more frequently called a structurally stable system) originally introduced by Andronov and Pontryagin, who formulated it as follows. For any $\varepsilon > 0$, there is a $\delta > 0$ such that for any perturbed system $\dot{\mathbf{p}} = \Phi_1(\mathbf{p})$ differing from the original one in terms of a given metric $|\Phi_1 - \Phi|$ by not more than δ, there is a one-to-one continuous phase space mapping that displaces its points by not more than ε and transforms the trajectories of the original system into those of the perturbed system.

Within one- and two-dimensional phase spaces, the so-called Morse–Smale systems are structurally stable. Their sets of non-wandering points include only finite numbers of stationary points and periodic orbits, all of them hyperbolic, while corresponding stable and unstable manifolds either never intersect or add up to form a complete tangent space (the property of transversality).

However, in multidimensional phase spaces, the Morse–Smale systems are no longer structurally stable, and for them to become rough it is necessary, as Smale hypothesized (1965), that each one-parameter phase-space transform possess a hyperbolic set Ω of nonwandering points, and in addition that the ensemble of periodic points be a dense covering subset of Ω (the so-called A-axiom). It is also necessary that each stable and each unstable manifold of points within Ω have the property of transversality. Ruelle and Bowen proved that the attractors of systems satisfying the A-axiom are stochastic. Thus, for a phase space with dimension greater than two, an infinite hyperbolic nonwandering point set Ω including a dense covering subset of periodic trajectories is indicative. There are remarkable phase flows, discovered by Anosov, for which the whole phase space is the hyperbolic set.

Another distinguishing feature of the strange attractor is that its Lebesque measure is zero. Hence, this subset (or, roughly speaking, volume) of the phase space actually has the structure of a Cantor set, i.e., a dense closed set with Lebesque measure zero. To identify the dimension of a Cantor set one uses a much finer measure, the Hausdorff measure.

The formation of a strange attractor can be pictured as follows. After a short period of time, an initially small phase-space cube starts expanding in the directions of instability and compressing in the directions of stability until it eventually takes the form of a long thin sheet, which then starts folding up due to the limited diameter of the attractor. This process continues until, in the limit of long periods of time, the initial phase-space cube becomes an infinitely thin film with an infinite number of folds. The Hausdorff measure of such a structure can be quite simply expressed in terms of the relative number of diverging and converging trajectories near a given point and the exponential rates of divergence and convergence.

The transition of a dynamical system into a stochastic state occurs through a sequence of bifurcations, i.e., dramatic changes in the pattern of motion caused by alteration of the system parameters. The most reliable hypotheses about the bifurcation sequences that give rise to stochastic events are called stochastization scenarios.

There are now several such scenarios that have been closely studied [2.5], although the agreement between the theory and the actual genesis of turbulence in fluids and gases is still open to question.

2.3 Weather Predictability

Let us now turn to the problem of weather predictability and estimate the terms over which forecasting is possible. The above excursion into dynamical-system stochastization theory was intended to help the reader understand the limitations on forecasting.

What we shall discuss now is long-term weather forecasting (several weeks ahead) within the theory of atmospheric dynamics. When we defined weather and climate we touched upon the division of motions into large- and small-scale ones. This division is very helpful in many areas of meteorology. Large-scale motions can be considered individually whereas small-scale motions can only be treated statistically. For example, when studying the distribution of chimney effluent in the air, a large-scale air motion, the wind, is predetermined individually, whereas a smaller-scale turbulence giving rise to sporadic volutes in the smoke current is given merely statistical treatment (e.g., by introducing exchange coefficients).

This approach is applicable to such areas of meteorology as long-term forecasting. But whereas large-scale weather characteristics are eligible individually for long-term prediction, smaller-scale motions, such as those producing smoke-current volutes, cannot and probably need not be forecast a long time ahead. Now it is only natural to ask where the dividing line between large-scale processes ('weather') that are forecast individually and small-scale processes ('turbulence') treated statistically is to be drawn.

The position of this line varies for different forecasting terms. Indeed, when a forecast is made for a very short period (e.g., less than 24 hours), we start tracing even mesoscale weather processes (using for this purpose weather-chart data drawn from networks of closely located stations); by contrast, weather processes generalized over years are more suitable for statistical treatment (or, as Defant suggested [2.6], are apt to be seen as macroturbulence).

Demarcation of scales implies choosing a distinct model from a set of models with different degrees of detail with respect to the phenomena under study. In this way, a hierarchy of models providing a hierarchy of degrees of determinacy [2.7] is established.

The fact that small-scale motions are practically unpredictable for long terms requires explanation. If we could accurately fix the initial state of any small-scale motion and accurately solve the relevant exact dynamical equations, then, abstractly, the terms of predictability would be unrestricted. Even the continuity of hydrodynamic fields would be no handicap to this, for within small lengths continuous fields feature linear variability, and it is sufficient to fix the initial state of the field using only a grid of points spaced closer than the internal scale of turbulence, which is of the order of several millimeters in the atmosphere. However, the initial states of meteorological fields can be fixed only on much coarser grids (with horizontal spacings 10^7–10^8 times as large as the internal scale). Thus, it should be readily apparent that individual motions on a scale smaller than the grid spacing are not fixed at all. Further uncertainties are introduced through measurements ('instrument noise') and truncating errors.

Therefore, even if the exact dynamical equations are solved correctly, the prediction is inevitably aberrant due to residual inaccuracy (as proved in Sect. 2.2), and the size of the aberration grows with time. Furthermore, the dynamical equations we use are actually approximations which get even more approximate when we reduce them to difference equations in order to obtain numerical solutions.

Clearly, prediction of individual processes provides information additional to that obtained by statistical ('climatic') computation, but only so long as prediction errors are not large enough to exceed mean climate-scale variations of predicted phenomena. The relevant term may be called the limit of predictability, and it evidently depends on the following:

– The types of the processes (specifically, their scales),
– The character and magnitude of initial residual errors,
– The accuracy of the forecasting method.

The determination of this limit is actually the predictability problem itself.

The idea that a small perturbation in initial atmospheric conditions (an initial residual error) can cause a significant variation in the final state at a later time, making prediction difficult, was put forward as long as 40 years ago by the prominent mathematician Kolmogorov. He expressed it figuratively like this. Imagine there are two almost identical planets with identical atmospheric conditions. If somebody steps onto the porch of a house on one planet and waves a handkerchief, whereas nobody does the same on the other planet, how much time will elapse before the weather conditions on the two planets become drastically different?

For a formalized predictability problem in meteorology, the size of a prediction error is commonly defined as the standard deviation of the predicted values from the true value. We shall use the same definition. Let $x_L(t)$ be the real state of an atmospheric component of scale L at instant t. Assume that the evolution of $x_L(t)$ is given by a dynamical system, and the true value is that computed using exact initial conditions. Because of the initial errors mentioned above, the predicted value $z_L(t)$ is different from the true value. Now, averaging the predicted values over the ensemble of all possible initial conditions (their distribution determined by a given probabilistic measure) we obtain the average predicted value $\overline{z_L(t)}$. The mean square prediction error is defined from

$$\sigma_L^2(t) = \overline{\rho^2\left[x_L(t), \overline{z_L(t)}\right]},$$

where the bar denotes an average over the ensemble of initial conditions. Distance ρ is the Euclidean distance for the case of finite-dimensional systems and the distance in the space of square integrable functions for the case of continuous systems. We also assume that the climate-scale probability distribution is prescribed for the multitude of possible atmospheric conditions, so that the 'climatic' variance of L-scale components is known:

$$\sigma_L^2 = \left\langle \rho^2 \left[x_L(t), \langle x_L(t) \rangle \right] \right\rangle,$$

where the angle brackets denote climate averaging (which is practically the same as averaging of long-term observational data). Then the predictability limit t_L of L-scale components is the upper bound of t values for which the condition $\sigma_L^2(t) < \sigma_L^2$ is met. This definition was first given almost 30 years ago [2.6] and is close to that of the recently proposed term 'predictability horizon' [2.7] (but not identical).

The problem of predictability in meteorology was first posed by Thompson [2.6], who also attempted to calculate the function $\sigma_L^2(t)$ analytically within the simplest (quasi-geostrophic) prognostic models. A more accurate computation was made by Novikov [2.6], who stated the following: if the basic initializing field and the initial error field are statically independent, homogeneous, isotropic, random fields with correlation radii L and L_1, we get

$$\sigma_L^2(t) = \frac{1}{16} \sigma_L^2 \left[1 + 2 \frac{(t/T)^2}{\left[1 + (L/L_1)^2 \right]^2} + \ldots \right],$$

where T is a time scale typical of weather forecasting variability. Given $T = 1$ day and $L = 2L_1$ we find that the time t_L in which the prediction error becomes as large as a random choice error equals two weeks. Equations following the evolution of the error field statistically were deduced by Tatarskii [2.6].

The first numerical experiment to look into predictability was staged by Diky and Koronatova [2.6], who compared 24-hour and 48-hour-ahead forecasting results using a barotropic model that contained 'true' initial data corresponding to a random residual error. They examined both a spatially uncorrelated error field and a field with a correlation radius of two grid spacings, and took the mean-square error as 1 and 2 decameters. Results differed for different series of forecasts, but on

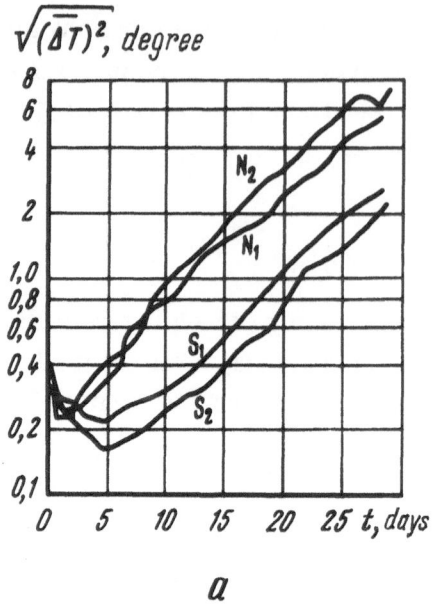

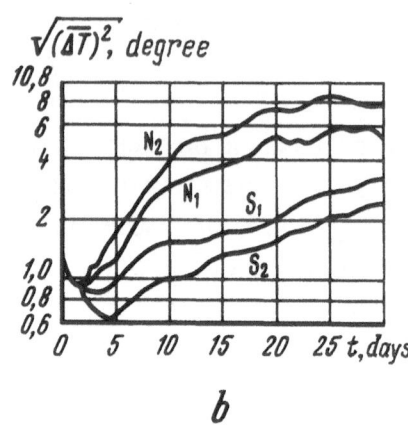

Fig. 2.1a. The rms error for air temperature forecasts made using the Mintz model [2.6] with the initial sinusoidal error $\delta T = \sin 6\lambda \cos 11\varphi$ (1° amplitude) introduced on the 234th day of the Mintz numerical experiment following the Charney procedure [2.6] (N is the Northern hemisphere, S the Southern hemisphere, 1 the ~400 mb level, and 2 the ~800 mb level)

Fig. 2.1b. The rms error for air temperature forecasts made using the Mintz model [2.6] with a random initial error (modulated with the $\cos \lambda \cos 6\varphi$ multiplier introduced by the Charney procedure)

average the errors grew 30–50% over 24 hours, the growth rate being somewhat higher for spatially correlated errors than for uncorrelated ones.

Far more complete numerical experiments concerning the predictability problem were run by Charney, who employed models devised by Smagorinsky, Leith, and Mintz [2.6]. Error variations were especially pronounced with the Mintz model, and it is worthwhile to consider them more closely. Initial errors were introduced in the temperature field, and the resulting errors were computed for two levels, level 1 of ~400 mb and level 2 of ~800 mb, each for the Northern (N) and Southern (S) hemispheres. Mean square errors of temperature forecasts for an initial sinusoidal error field $\delta T = \sin 6\lambda \cos 11\varphi$ with an amplitude of 1° (where φ is a latitude and λ a longitude) are illustrated in Fig. 2.1a. Shown in

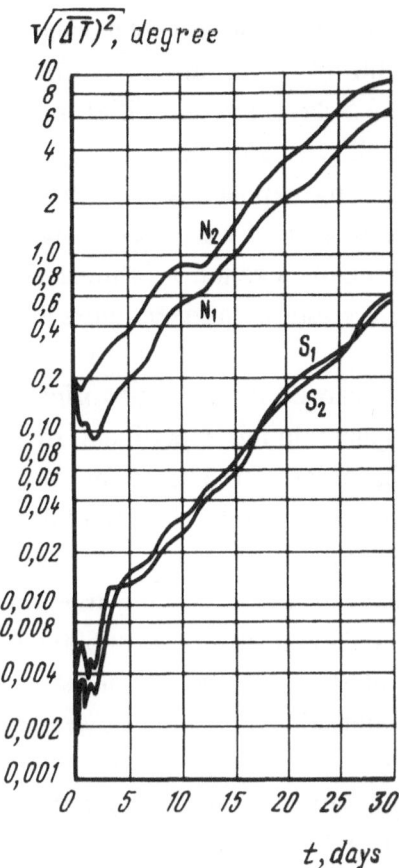

$\sqrt{(\overline{\Delta T})^2}$, degree

t, days

Fig. 2.2. The rms error for air temperature forecasts made using the Mintz model with an initial error localized by the Charney method [2.6] in regions 21–63°N, 157–203°S (note that the gravitational waves induced in the Northern hemisphere by a disturbance due to adjustment of velocity and pressure fields reach the Southern hemisphere in 1–2 days, giving rise to disturbances whose energy is 3 orders of magnitude lower than that of the initial ones)

Fig. 2.1b is the evolution of the prediction error for the case where a random initial error is introduced, while the evolution for the case of an initial error that is localized in space (the square within 21–63°N, 157–203°S) is displayed in Fig. 2.2.

Charney's limit of predictability is the time in which a prediction error reaches a mean square difference between two random temperature fields (or two real fields: if the time difference is more than 3 days, the mean square difference is almost constant, that is, 5°C in the Northern and 4°C in the Southern hemisphere at level 1, and 8°C and 3°C respectively at level 2, which suggests among other things that a temperature field can be predicted inertially for no longer than 3 days). With the difference of 8°C (at the 800-mb level in the Northern hemisphere), an error doubling period of 5 days, and an initial mean-square error of 1°C, the limit of predictability is $5\log_2 8 = 15$ days (the fact that errors of different origin have the same exponential growth rates is explained as the effect of baroclinic instability of waves with a wave number $m = 6$ along latitudes that are readily excited at almost any spectrum of initial disturbances due to a strong nonlinear interaction between spectral components).

Another experiment of this kind was carried out by Smagorinsky, who used a nine-level model of the atmosphere. In this experiment, small random uncorrelated disturbances of the order of 0.5°C in the mean temperature of an air column were introduced into true initial field data. In 14 days this resulted in relatively small-scale prediction errors, so that the correlation coefficient for fields that were forecast using true and distorted initial data fell within 14 days to 0.99 at 50 mb, to 96 at 500 mb, and to only 0.88 at 1000 mb. Using the pattern of error growth at the end of this period as reference, Smagorinsky inferred that the limit of weather predictability within this model was 3–4 weeks. The larger predictability range in comparison with that deduced by Charney is probably due to the considerably greater number of degrees of freedom in this atmospheric model. If so, then limits of predictability computed by modeling must be lower estimates, since the number of degrees of freedom in any model is much less than that for the real atmosphere.

During the years that followed, the concept of predictability limits in meteorology remained practically unaltered, despite the availability of powerful computers and advances in theoretical science [2.3], and in none of the models was a term of one month achieved. Still, researchers came to understand the reason for their slow progress, namely the sto-chastization of nonlinear dynamical systems representing the atmosphere which we discussed in Sect. 2.2.

Determination of predictability limits is not an end in itself. A constructive solution of the predictability problem for one or another long term is in effect the determination of which meteorological field characteristics are predictable for a given term. If individual synoptic processes, such as cyclones or anticyclones, were unpredictable more than 2, 3, or 4 weeks in advance, say, then this would not necessarily mean that other weather characteristics could not be forecast over similar time scales. In fact, there are generalized (averaged) characteristics of ensembles of individual processes, for example, charts of monthly rainfall sums, that are of great practical interest. However, the discussion of monthly averaging is beyond the scope of this section, which is on weather and not climate, so we shall postpone it for a while.

2.4 Elements of Stationary Random Process Prediction Theory

Let us assume that the behavior of a physical quantity x (taken to be scalar for simplicity) of interest to us is given by a random process $x(t)$, and that we are therefore concerned only to find the probability that the value of $x(t)$ falls into one or another interval of values. The set of joint probabilities corresponding to all possible sets of times is called the set of finite-dimensional distributions. To describe a random process fully, one has to prescribe all its finite-dimensional distributions.

Suppose that observations $y(t)$ of the process $x(t)$ continue until instant t. In the general case, $y(t)$ does not coincide with $x(t)$ because of measurement errors. Let us denote the set of all $y(t)$ values that took place up to time t by Y_0. Thus all information on the behavior of the process in the past is contained in the set Y_0. The problem of making an optimal forecast for term τ is formulated in accordance with the mean-square error criterion adopted above as follows: determine quantity $z(t_0, \tau)$ depending only on Y_0 variables for which the standard deviation

$$\sigma_0(\tau) = \left(\left\langle \left(z(t_0, \tau) - y(t_0 + \tau) \right)^2 \right\rangle \right)^{\frac{1}{2}}$$

is minimal. Here the angle brackets denote the average over the ensemble of $y(t)$ realizations. The required quantity $z(t_0, \tau)$ is a conditional expectation (mean) value of quantity $y(t_0 + \tau)$ relative to the set Y_0 of random variables and can in the general case be presented in terms of finite-dimensional distributions of processes $x(t)$ and $y(t)$. However, the practical importance of this expression is small because of its awkwardness and because finite-dimensional distributions are actually unknown to experimentors.

To simplify the problem, we shall restrict ourselves to the class z of quantities that are linearly dependent on Y_0 variables. Clearly, this restriction sharply reduces the credibility of the prediction. The situation is as if we agreed to approximate real dynamical systems to linear ones.

By way of example, consider a case illustrating the magnitude of the possible error. Assume that x is a random quantity with a zero mean, and that the probability that x takes values larger than α equals the probability that x takes values smaller than $-\alpha$. Observational errors are presumed to be absent. The task is to furnish the best possible linear estimate (a forecast) of quantity $u = x^2 - \langle x^2 \rangle$, i.e., to find a coefficient k providing a minimal standard deviation of $z = kx$ from u. In other words, the problem is to determine regression z over x. Using a standard formula expressing the fact of noncorrelation between $u - kx$ and x we obtain

$$k = \frac{\langle ux \rangle}{\langle x^2 \rangle} = 0.$$

Such a paradoxical result, indicating that an accurately precomputed value is predicted to be zero, is due to replacement of the notion of dependence by that of correlation.

Still, there is a fairly extensive class of processes for which the best possible linear forecast coincides with the one that is best in general – these are the Gaussian processes [2.17]. As we shall see later, a climatic system can, in rather general terms, be considered as a linear system, and therefore application of a linear prediction theory to climate investigation is justifiable.

Let us introduce a scalar product into the space of random quantities (presuming, without any loss of generality, that $\langle x(t) \rangle = \langle y(t) \rangle \equiv 0$) expressed by

$$\left(y(t_1), y(t_2) \right) = \langle y(t_1) y(t_2) \rangle.$$

The scalar product introduced here converts the set of random quantities composed of the process values at various points, their linear combinations, and limits of these combinations in conformity with norm $\| \cdot \| = (.,.)^{1/2}$ into Hilbert space L. For example, if process $y(t)$ is generated by a finite number of harmonics, the corresponding space L is finite-dimensional.

Let L_0 be a linear subspace of L generated by just the random quantities from the set Y_0. In the finite-dimensional case, L_0 is a hyperplane. Now the problem is to find a point in hyperplane L_0 closest to point $x(t_0 + \tau)$. To do this requires 'dropping a perpendicular' from point $x(t_0 + \tau)$ onto hyperplane L_0. This job is done at school using a ruler and a pair of compasses, but those tools are insufficient for our purposes. However, the geometric approach to the problem, relating to one of the fields of functional analysis rather than to random process theory, is very helpful. The mean-square error of an optimal forecast $\sigma_0^2(\tau)$ is actually the length of the perpendicular we need, while the required predictand is its base, denoted as $z(t_0, \tau)$.

The prediction problem has a relatively complete solution in the event of a statistically stationary random process. Here we consider processes that are stationary in the broad sense, i.e., with the correlation function $R(t_1 - t_2) = \langle x(t_1)x(t_2) \rangle$ depending only on the difference of time moments. Observational errors are neglected, i.e., $x(t)$ is identical with $y(t)$. Under rather general conditions that show themselves in the form of a continuous spectrum of the $x(t)$ process, the latter can be expressed in the form of Vold's expansion:

$$x(t) = \int_{-\infty}^{t} G(t-s)b(s)\,ds,$$

where $b(s)$ is a stationary white noise with unit variance, and the kernel $G(u)$ is determined by the correlation function. The solution of the prediction problem is given by:

$$z(t_0, \tau) = \int_{-\infty}^{t_0} G(\tau-u)b(u)\,du.$$

Thus an optimal linear prediction can be made if one knows the correlation function of the process. For example, consider a random process with correlation function $R(t) = \exp(-\alpha t)$, where $\alpha > 0$ is a parameter. With the additional assumption of a Gaussian distribution this process can be regarded as a Markovian one, known in the mathematical literature as the Ornstein–Uhlenbeck process. It is also known to physicists as the solution of the Langevin equation:

$$\dot{x} + \alpha x = b(t).$$

In the latter case, $G(u) = \exp(-\alpha u)$, and consequently $z(0, \tau) = \exp(-\alpha t)\, x(0)$, which suggests that the prediction depends only on the last observable value – a characteristic Markovian feature. The prediction error can be easily determined too: $\sigma^2(\tau) = 1 - \exp(-2\alpha\tau)$. Note that the equation $\sigma^2(\tau) = \gamma$ defining the limit of predictability is solved, given the preset prediction accuracy γ, by the formula

$$\tau_\gamma = -\frac{1}{2\alpha}\ln(1-\gamma).$$

A faithful analytical solution of the problem is available for a more general case in which the spectral density of the process equals unity divided by a polynomial of degree $2p$. Here an optimal prediction for any desired term is expressed by the value of the process and its derivatives up to and including order $p-1$. Such processes are called p-Markovian and are solutions of stochastic differential equations with constant coefficients of order p and with white noise on the right-hand side. Also, the prediction problem has an explicit solution in the case when the spectral density of the process is a rational function [2.9]. Such processes are called complex-Markovian and are characterized by the ability to integrate into vector Markovian processes as components. Complex-Markovian processes also satisfy stochastic differential equations with constant coefficients; however, not only white noise but also its derivatives can be found on their right-hand side. The set of predictors for a complex-Markovian process is also finite-dimensional, but not all the predictors coincide with the derivatives at the 'present' moment of time – some are functions of the entire prior process [2.10].

An analogue of the complex-Markovian process in the discrete-time case is the so-called autoregression/moving-average (ARMA) process now extensively employed for modeling many natural processes, including those related to climatology. The evolution $\{x_n\}$ of such processes is given by the difference equation

$$x_n + \alpha_1 x_{n-1} + \ldots + \alpha_{p-1} x_{n-p+1} = \varepsilon_n + \beta_1 \varepsilon_{n-1} + \ldots + \beta_q \varepsilon_{n-q}, \qquad (2.5)$$

where $\alpha_1, \ldots, \alpha_{p-1}$ are constant coefficients of autoregression, $\beta_1, \ldots, \beta_q$ are constant coefficients of the moving average, and $\{\varepsilon_n\}$ is discrete white noise. The pair (p,q) is called the process order. When all the coefficients β_i equal zero, the process is simply an autoregression.

The first-order autoregression process ($p=1$) is very important in terms of applicability. Its optimal prediction (in the absence of observational errors) is given by the simple formula $z_{n+1} = \alpha_1 x_n$. Notice again that if $x(t)$ is an Ornstein–Uhlenbeck process, then sequence $\{x_n\}$ derived from $x(t)$ by choosing values separated by equal time spans is the first-order autoregression process.

In dealing with ARMA processes given by equation (2.5), one can adopt the well-developed technique of computing autoregression and moving-average coefficients using observational data; furthermore, explicit prediction formulas are available [2.11]. It should be noted that if $x(t)$ is a complex-Markovian process then the sequence $\{x(n\Delta t)\}$, where

Δt is a given time increment, is an ARMA process. However, if $x(t)$ is a p-Markovian process with a 'unity-divided-by-a-polynomial' spectral density, $\{x(n\Delta t)\}$ is not necessarily an AR process. In this way, a relationship is established between stochastic differential equations appearing in different physical problems and ARMA models introduced as a means of processing time series of unspecified origin. Such a relationship is found to exist between stochastic differential equations with partial derivatives and autoregression vector processes. Because it is complicated, however, this relationship will not be discussed here. The relevant formulas and calculations are given in the monograph by Piterbarg [2.12].

The results obtained prove to be of great utility in constructing stochastic models of processes on the basis of physical models described by corresponding differential equations. In this way, it is also possible to estimate unknown or indefinite parameters in the physical model by using observational data along with a well-developed statistical technique for determining unknown autoregression coefficients. This approach was used in studies of climatic variability [2.12], the topic we now turn to.

2.5 Predictability of Climatic Processes

The conception of climatic variability developed by Hasselmann [2.13] and presented here is not widely accepted in climatology. The basic idea is that long-period (longer than a month) oscillations occurring within the ocean–cryosphere–land–atmosphere system are actually the response of a linear inertial system to short-period weather disturbances caused by synoptic-scale atmospheric processes and regarded as 'white noise'. Following this conception, the causes of climatic variability are to be found within the climatic system itself rather than among various external parameters (for example, those associated with variations in solar activity), although the latter should be accounted for. The advantage of this approach is that it enables one to study the predictability problem. Roughly speaking, all simple stochastic climatic models can be reduced to a version of the Langevin equation where the white noise represents weather-scale atmospheric processes and the state variable represents some climatic-system parameter, such as the ocean surface temperature, thickness of the ice sheet, or land vegetation. Another convenient feature of this approach is that we need not look for the cause of random climatic variations (their existence is beyond doubt), for it is known to lie in the stochasticity of atmospheric behavior, which stems in turn from the nonlinearity of the underlying dynamical processes. This is the shortest path to explaining the irregularity of climatic variation evident from, for example, the fact that spectra of long-period fluctuations of hydrometeorological characteristics take the form of 'red noise' [2.14]. Furthermore, it would be incorrect to say that climatic variability is commonly associated with a time scale exceeding one month. We shall now present the concepts behind simple stochastic climatic models using Hasselmann's terminology [2.13].

At each instant of time, the state of a climatic system is given by a set of fields $W = \{w_1(\mathbf{r},t), w_2(\mathbf{r},t),...\}$, where t is time and $\mathbf{r}$ is a point in space. The set W can be divided into two subsets, one including slowly varying (climate-scale) variables $U = \{u_1, u_2,...\}$ and the other comprising rapidly varying (weather-scale) variables $V = \{v_1, v_2,...\}$. The V group contains not only atmospheric fields but also a number of oceanic fields whose variability is directly induced by synoptic processes, such as abnormal drift currents, seasonal thermocline depth, etc.

The temporal characteristics of group V variables fall within a few days (an atmospheric weather scale), while for group U variables, such as mean sea-water temperature in the upper mixed layer, scales of several months and more are typical. Therefore, the temporal scales τ_u and τ_v for the variability that exists are related by:

$$O\left(V_i\left(\frac{\partial V_i}{\partial t}\right)^{-1}\right) = \tau_v \ll \tau_u = O\left(U_i\left(\frac{\partial U_i}{\partial t}\right)^{-1}\right).$$

The difference between τ_u and τ_v is due to the difference in fundamental physical characteristics between water and air, i.e., specific heat, mass density, and kinematic viscosity. The evolution of a climatic system may be written:

$$\dot{u}_i = P_i(V,U) \tag{2.6}$$
$$\dot{v}_i = Q_i(V,U), \tag{2.7}$$

where P_i, Q_i are nonlinear operators for the multicomponent functions V, U in the general case. Equation (2.7) is the general atmospheric circulation (GAC) model, while equation (2.6) represents the variability of the sea surface temperature and other slowly varying climatic characteristics as a response to 'random' excitation by weather variables. Given the present state of the art in computer engineering, integration of equation (2.7) over times of the order of τ_u is impossible. A higher-resolution model is usually employed for integration of equation (2.7) over an intermediate period τ_0, $\tau_v \ll \tau_0 \ll \tau_u$, during which slow variables can be considered as constants, but which is long enough for the collection of statistical data for weather-scale variables. Hence, though the GAC models provide important information for climatic variability studies, they fail to simulate this variability.

Traditionally, an investigation into the dynamics of long-period variability proceeds by averaging equation (2.6) over the τ_0 period. The resulting statistical dynamical model is:

$$\frac{\partial}{\partial t}\langle u_i \rangle = \langle P_i(\mathbf{V}, \mathbf{U}) \rangle. \tag{2.8}$$

Formally, the $\langle \ \rangle$ operation in equation (2.8) is more suitable for averaging over an ensemble of rapidly varying realizations of the $\mathbf{V}$ field. With the assumption of ergodicity, this averaging accords very closely with time averaging. Since operators P_i are nonlinear in the general case, the average variability rate $\langle P_i \rangle$ of the $\langle u_i \rangle$ parameter depends upon the statistical characteristics of $\mathbf{V}$ and $\mathbf{U}$, taken both individually and in combination. In order to bring the equations to closed form, we must now state some hypotheses. The need to do so indicates the imperfection of the static dynamical models employed. Although equation (2.8) is closed with statistical models, the reduced equation is actually deterministic, not statistical, due to averaging and the use of hypotheses allowing it to be brought to closed form. As mentioned above, the asymptotic solutions to nonlinear deterministic equations involving a relatively small number of degrees of freedom suffice to represent nonperiodic oscillations of a random type similar to weather-scale or climate-scale fluctuations [2.15]. In most cases, however, the well-known simple statistical dynamical models predetermine a unique time-independent asymptotic state for any given initial condition. Intrinsically, such models cannot provide the continuous-spectrum solutions found in nature. In the past, the long-period variability occurring within classical statistical dynamical models (SDMs) was regarded as the response of system (2.8) to variations of external boundary conditions, such as solar radiation and atmospheric turbidity, rather than a result of interactions taking place inside the system.

However, if instead of averaging equation (2.6) to give system (2.8), we treat it as a system of stochastic differential equations, we will be able to define the causes of long-period variability directly from internal interactions (which, of course, does not eliminate the possible significance of additional climatic variations induced externally). In using this approach, slow climate variables should be looked upon as random fields with time-correlation scales of τ_u and larger. Specifically, a long-period variability time-scale representative of the 'red noise' can be obtained from equation (2.6) using the following logic.

Functional P in equation (2.6) generally depends on the past behavior as a whole: $V_t = \{v(\mathbf{r},s), s \le t\}$, $U_t = \{u(\mathbf{r},s), s \le t\}$. Assuming deviations $v' = v - \bar{v}$, $u' = u - \bar{u}$ from the mean value to be small, we linearize equation (2.6):

$$\dot{u}' = \frac{\delta P}{\delta V_t} v' + \frac{\delta P}{\delta U_t} u', \tag{2.9}$$

where $\delta P/\delta V_t$, $\delta P/\delta U_t$ are variational derivatives at point $(\bar{V}_t, \bar{U}_t)$ which are known to be linear operators of v' and u' respectively. Under the simplest assumption that these operators are dependent merely on the present state, we obtain the Langevin equation from equation (2.9):

$$\dot{u}' = \alpha u' + \sigma v',$$

where α and σ are constants. With allowance made for a small correlation time for the 'fast' process $v'(t)$, this equation provides the 'red' spectrum for $u'(t)$.

Note that it is not necessary to linearize equation (2.6). The assumption of a short correlation time makes it possible to study the statistical properties of solutions of the heat balance equation, since equations for the first and second statistical moments of the solutions are closed and linear [2.12]. Hasselmann called the unaveraged models of the type described by equation (2.6) the models of stochastic generation [2.13]. According to him, the relationship between the statistical dynamical model, the GAC models, and models of stochastic generation is established using an analogy with Brownian motion. Climate variables U and weather variables V may be interpreted in terms of Brownian motion as large and small particles respectively, or rather as their position and momentum coordinates.

In terms of this analogy, the analysis of climate variability within the SDM is equivalent to determining the trajectories of large particles solely by means of the interactions of these particles with each other and with the mean fields of pressure and stress created by the movement of smaller particles (plus the effect of external forces). Continuing the analogy, numerical experiments with the GAC model are equivalent to an accurate computation of all the trajectories of the small particles, given fixed positions for the large ones. Even if the large particles change their positions during the experiment, it is usually impossible to perform an integration over a period that is long enough to allow substantial deviations of the larger particles from their original positions. Finally, the stochastic generation models are in line with the classical statistical conception of Brownian motion, whereby the laws governing the fluctuations of the large particles are deduced from the statistics of the small particles with which the large ones interact.

In contrast to the Brownian motion analogy, the variables v_i associated with a real weather–climate system are naturally out of thermodynamic equilibrium, and their statistical properties cannot be derived from the statistical thermodynamic theory of energy-conserving systems. However, these properties can be obtained from the results of numerical experiments or from observational data. The statistical approach offers a substantial saving in time and labor, since the investigator needs only a relatively small body of statistical information concerning V. This can be derived from the GAC model experiments using relatively short time spans, $\tau_0 \ll \tau_u$.

It may seem strange at first sight that the statistical reduction of the entire weather–climate system requires no closure hypotheses. Indeed, any expert familiar with turbulent fluid flow systems knows very well that their evolution is bound up with strongly nonlinear irreversible processes. In our case, however, the reduction is the result of a separation of temporal scales. In ordinary turbulent systems this feature becomes lost. For example, it is impossible to construct a simple quasi-stationary stochastic generation model for the atmosphere that can be associated with a model for the ocean. One of the main problems that precludes this is the derivation of linear transfer functions for the atmospheric response which would account for nonlinear eddy flow fluctuations.

Hasselmann's idea of simple stochastic models can be regarded as an alternative to the averaging concept that was dominant until quite recently in statistical physics. In the broad sense, that concept is implicit in the idea that large-scale processes can be described by deterministic equations in which random small-scale fluctuations are taken into account by means of effective coefficients that depend on the statistical characteristics of the 'microworld'.

Experiments to determine the statistical predictability of climate were initiated by Privalskii in the 1970s [2.16, 2.17], and the results were in general as discouraging as those reported later by Hasselmann [2.13]. Hasselmann's results were obtained from a theoretical analysis based on the Langevin equation. Privalskii's work [2.16–2.18] was concerned with the problem of forecasting the sequence of annual mean sea-surface temperatures (SST) in the North Atlantic over the period 1881–1970. The series were fitted using the ARMA models and employed an optimization procedure for picking the model's order (p, q). In most cases, the first-order autoregressive model, AR(1), proved to be optimal, while the relative accuracy of one-step-ahead (one-year-ahead) forecasts was 75–80%. In cases where some other model appeared optimal, the predictability it provided was practically the same as that obtained using AR(1). These results were in full agreement with what the simple stochastic model implied. Privalskii assumed the predictability limit to be the time in which the relative prediction error $\rho(\tau) = \sigma^2(\tau)/\sigma^2$, where σ^2 was climatic variance, reached 80%. For annual average SSTs this limit is consequently one year.

Predictability estimates of mean monthly SST anomaly series look more optimistic [2.12]. By 'anomalies' here we mean monthly temperature deviations from the norm calculated for lengthy series of observational data. Over the most part of the North Atlantic the mean-square prediction error ρ for a one-step-ahead forecast was 50–60%, which corresponds to a predictability limit of 3–4 months. In some regions predictability was appreciably higher or lower than the mean value owing to specific hydrological properties of these regions. Note that the period of 3–4 months was known to be the approximate duration of anomalies long before their predictability was investigated. For the northern part of the Pacific Ocean, because longer observational series were available (32 years as compared to 18 years for the Atlantic), it was possible to estimate the predictability of temperature anomalies individually for summer (May, June, July, and August) and winter (November, December, January, and February) months. Predictability was rather low for summer months (like the average predictability of seasonal temperature for the Atlantic), but in winter ρ went down to 30–50%, corresponding to a predictability limit of 6–10 steps. This is probably due to the fact that in winter the winds are stronger than in summer, and the upper mixed layer is thicker and consequently more inert.

These results were obtained using 'multidimensional' (vector) AR models, where the 'state' data were temperature data transferred onto a regular 5° grid along with corresponding SST values specified for four neighboring grid points. This allowed spatial relations within the temperature field (the advection and large-scale diffusion of heat) to be taken into account. But the gain in predictability was no more than 5–10% compared to one-dimensional models. Moreover, a simultaneous statistical analysis of annual mean temperature series based on the data mentioned above that had revealed a feedback between the systems of warm and cold currents in the Atlantic provided practically the same insignificant gain in predictability [2.18]. The high prediction errors were caused, among other things, by a considerable margin of error in the experimental data. Their estimated 'noise' component was about 20% in terms of variance. This can be attributed to the heterogeneity of the data, which was collected from passing merchant ships. The potential predictability on the assumption of a negligible noise component was estimated using a special noise-filtering procedure [2.12]. The lower limit of the prediction error ρ proved to be 20–50% depending on the location in the North Atlantic. These results give an idea of the predictability horizon of the SST anomalies.

The notion of a 'predictability horizon' was originally introduced by Lighthill and further specified by Kravtsov [2.7], who defined it as the time τ_{lim} over which the correlation between the measured value $y(t_0 + \tau)$ of process $x(t)$ and the expected value $z(t_0 + \tau)$ reaches 0.5 in the absence of observational errors and provided an ideal (adequate) model is used. The same term was used, though in a somewhat different context, by Parsen and Newton [2.19]. Since the correlation coefficient $D(\tau)$ is expressed in terms that relate it one-to-one with the mean-square prediction error $\rho(\tau)$, which in the AR(1) case equals ρ^τ, where ρ is the one-step-ahead prediction error, it is reasonable to consider the predictability horizon as a function of ρ. At this point we mention that the statistical model of SST anomalies employed here was derived from the evolutional equation of the heat balance of the upper mixed ocean layer, from which we conclude that the model can be considered ideal (i.e., an adequate represention of the actual processes). For a minimal $\rho = 20\%$, the predictability horizon τ_{lim} is approximately 12 months, while for $\rho = 50\%$, $\tau_{lim} \cong 4$ months. These figures follow from the simple relationship $D(\tau) = 1 - \rho(\tau)$ which is valid for the AR(1) case.

An important quantity for oceanographers is the sea level. Results obtained by processing annual mean sea-level series of length 70–100 years, based on data acquired from 35 onshore stations, are summarized in [2.18]. The models that proved optimal were AR models of orders $p = 0$ ('white noise'), 1 and 2, with ρ varying within 70–80% for level fluctuations due to monsoon surges and reaching 90% for fluctuations associated with zonal atmospheric circulation. Strictly speaking, it might be more correct to use the term 'unpredictability' rather than 'predictability' here, for even by the most unassuming standards the limit of predictability is hardly greater – and sometimes smaller – than the discreteness step.

The outcome of studying the predictability of annual mean air temperatures representing the mentioned close correlation between local atmospheric processes was even more disappointing [2.18]. Nearly 100 series of length 80–315 years were processed. A one-step ahead prediction error ρ was within 85–99% for optimal AR models of orders $p = 0 \div 7$. The limit of predictability was less than one year.

A search for a higher statistical probability led us to the conclusion that averaging atmospheric and oceanic processes in space is likely to increase it. For example, the error of a one-step-ahead forecast of the annual mean air temperature for the Northern hemisphere was about 60%, and the predictability limit reached 5 years.

On averaging annual air temperatures within 5° latitudinal zones in the Northern hemisphere (from 35°N to 75°N) over the period 1891–1976, ρ proved to be 60–70%, while the corresponding limits of predictability were 4–5 years. This result is probably due to lessening of meridional heat and potential energy transport in those zones. For middle and low latitudes, where the meridional circulation is intense, the limit of predictability is about one year. Estimation of mean zonal air pressure predictability yields similar results [2.18].

Further, for even the relatively small mean-square error of a one-step ahead air temperature forecast, a period for reliable forecasting corresponding to a probability of 0.9 of being true is too large, since it considerably exceeds a double standard deviation computed over the whole observational data series.

We find an improved level of statistical predictability in the case of one particular spatially-averaged oceanic process, namely, changing ice-cover area in the Barents Sea in the summer months of 1899–1981. The relative one-step ahead prediction error for this case is only 53%, and the limit of predictability is 3 years [2.18].

In striking contrast with these figures are the results of research into the predictability of enclosed lake levels. The method of estimating predictability employed in this work was entirely different. Privalskii [2.18] proceeded from the second-order autoregression model derived directly from the water balance equation, which in a simplified form is written as

$$x_t = \alpha_1 x_{t-1} + \alpha_2 v_t + \alpha_3 l_t , \qquad (2.10)$$

where x_t is the lake level in the t-th year, v_t is the annual inflow volume, l_t is the annual thickness of the apparent vaporization layer, $\alpha_1 = 1 - b_1 < e$, $\alpha_2 = 1/\overline{F}$, $\alpha_3 = -1$, $\overline{F}$ is the average lake area, and b_1 is the proportionality factor for the regression relationship $\overline{F} = a + b\overline{w}$, where $\overline{w}$ is the average lake volume. Using models of apparent vaporization and inflow constructed from observational data, we obtain the following figures for the autoregression coefficient a_1: Caspian Sea 0.97, Lake Balkhash 0.91, Dead Sea 0.98, Great Salt Lake 0.87. These estimates imply a rather high predictability limit for the level of an enclosed lake – from 10 years for the Great Salt Lake to 50 years for the Dead Sea (the steeper the shore slopes and the less the mean inflow the better the predictability).

The high predictability is probably due to lags in level fluctuation that are typical of large enclosed lakes. Results obtained from measurement data were in good agreement with predictability estimates based on the above model [2.6].

2.6 Ways to Improve Statistical Forecasting

From the examples discussed above, we can conclude that averaging certain hydrometeorological characteristics over time and space usually results in an improvement in their predictability. This is the case, in particular, with air temperature. The same was observed with sea surface temperatures [2.20], where the limit of predictability was raised considerably by smoothing temperature anomalies over a three-month period. However, this rule lacks the power of generality. There are many examples to show that averaging observational data over an area yields no improvement in forecasts [2.18].

It seems reasonable to assume that employing more and more predictors allows us to achieve ever more accurate forecasting results. For example, sea surface temperature anomalies seem to be more predictable when we consider data not only for neighboring points but also for remote ones, and when we use data on air temperature and humidity, wind velocity, and so on, for all these factors are undoubtedly involved in the determination of water temperature variations at a given point. Yet simple numerical experiments in which the number of predictors was increased from 5 to 9 proved that this is not exactly so: not only did predictability remain generally the same, but in some cases it even deteriorated. This situation was analysed in detail by Yaglom [2.21]. Let us consider some of his remarks.

Suppose that we want to use the least-square method to obtain the best possible estimate of the predictand x using predictors $y_1, \ldots, y_k$, that is, to derive coefficients $a_1, \ldots, a_k$ that provide the least mean square of deviation $\langle (x - a_1 y_1 - \ldots - a_k y_k)^2 \rangle$. The empirical values of correlation coefficients η_i and η_{ij} between quantities x and y_i, $i = 1, \ldots, k$, and between pairs y_i and y_j, $i, j = 1, \ldots, k$, are not exact; they depend on the quantity (and quality!) of available data. Consequently, the optimal values $a_1, \ldots, a_k$ are also inexact and cannot be considered as the best for the entire general set of random quantities $(y_1, \ldots, y_k, x)$. They would be far less effective if they were applied to a subsequent independent sample. To avoid this complication, the collection of predictors must comply with several requirements, the most important of which is that in order to provide a sufficiently accurate evaluation of regression coefficients, the number of predictors k must not be too large. If the sampling size is of the order of a hundred or several hundred, k must not exceed several units. Lorenz called this constraint on the number of predictors 'the statistical prediction taboo' [2.22].

Another requirement initially pointed out by Kolmogorov [2.33] is the prohibition on exhausting a large number of predictor systems (even though each of them may contain only a few quantities) to select the best of them. The point is that when we look at a large number of different predictor systems we come across at least one system whose correlation coefficient between predictors and quantity x is much overrated. It is this system of predictors that should be chosen as the best. Yet already in the next experiment the chosen system of predictors yields results that are much worse than expected.

As Yaglom noted [2.21], Kolmogorov's considerations undermine the validity of several new methods of statistical weather forecasting. Among these, the screening method is typical. It goes as follows. Initially, a large collection of predictors is considered, from which the 'best' are selected in order. The first selected predictor (call it y_1) must fit the largest empirical correlation coefficient with x, the second (y_2) must be the main contributor to the combined correlation coefficient for the pair (y_1, y_2) and x, and so on. The screening procedure can be stopped after a few steps, since the remaining predictors make a practically negligible contribution. However, the above arguments lead us to believe that the 'best' collection of predictors can be obtained only for a particular sample, while for another, independent sample it may prove inadequate. The situation is practically the same as with the commonly used method of empirical orthogonal functions [2.24], which is in many ways similar to the case where the basic components of multidimensional statistical analysis are used [2.25]. Though commonly applied to the forecasting of oceanographic [2.20] and meteorological [2.22] fields, the screening method yields results that are not always satisfactory for cases where the 'higher' empirical modes are selected using screening-like methods [2.22].

One of the more helpful approaches for improving forecasting results is to choose, from a set of functions, the function of the quantities under study which gives the best predictions. Formally, the problem is formulated as follows. Given a set of observational data $Y = (y_1,...,y_n)$, derive the function $F(x_1,...,x_m)$ of random values $x_1,...,x_m$ whose optimal mean-square forecast based on Y is the best among all the forecasts from a set of functions F. If the problem is stated for a set of linear functions, it can be solved using a procedure proposed by Obukhov and Hotelling [2.26, 2.27]. According to them, for the given set of random values, one can find corresponding linear combinations u_i, $i = 1,...,n$, and v_j, $j = 1,...,m$, of values y'_k and x'_l respectively (the primes denote deviation from the mean value), such that both the u_i and v_j values appear to be uncorrelated, both internally and with each other, for $i \neq j$. Let $\rho_1 \geq \rho_2 \geq \ldots$ denote correlation coefficients for u_1 and v_1, u_2 and v_2, etc. In such a case, $\rho_1, \rho_2, \ldots$ are called the canonical correlations of sets Y and X, and random variables u_1 and v_1 the corresponding canonical variables. Clearly, value v_1 provides the solution to the problem. The derivation of the canonical correlations and variables can be reduced to an algebraic (matrix) eigenvalue problem.

Examples illustrating the application of the canonical correlation method to the case where Y is a set of observational data on atmospheric pressure obtained at a given network of stations on a given day are presented in [2.28], where it is shown that using the data for the day prior to observation one could forecast certain averaged characteristics appreciably better than pressure values for a given point. Here the general character of 'the best predictable properties of the pressure field' (fitting the first and second canonical correlations) is almost identical for the European part of the former USSR and for the USA (Fig. 2.3).

Another interesting example of the application of the canonical correlation method is to the series of actual mean temperatures on the Earth's surface averaged over a latitudinal zone within 30°N and 80°N [2.28]. Using climatic data records for 1892–1976, time correlation functions were determined for a sample size of 86 and used later to compute the first canonical variables u_1 and v_1 for sets $\{T(t), T(t-1), \ldots, T(t-n+1)\}$ and $\{T(t+1), T(t+2), \ldots, T(t+n)\}$ at $n = 1, 2, 3, 4, 5$. Correlation ρ_1 was found to increase monotonically from 0.6 to 0.8 with growing n. It was also found that using values $T(t), T(t-1), \ldots, T(t-4)$ one can do better at predicting the linear combinations of $T(t+1)$, $T(t+2), \ldots, T(t+5)$ that are close to the arithmetic mean of these values than the $T(t+1)$ value alone. This result is in good agreement with the hypothesis cited above that averaged characteristics are more predictable.

The canonical correlation method has also been worked out for the case of continuous sets $\{x_t\}, \{y_t\}$ [2.29]. With the development of these methods we obtain not only an elegant mathematical theory but also a helpful tool for tackling meteorological problems where it is important to trace and forecast the time behavior of atmospheric fields.

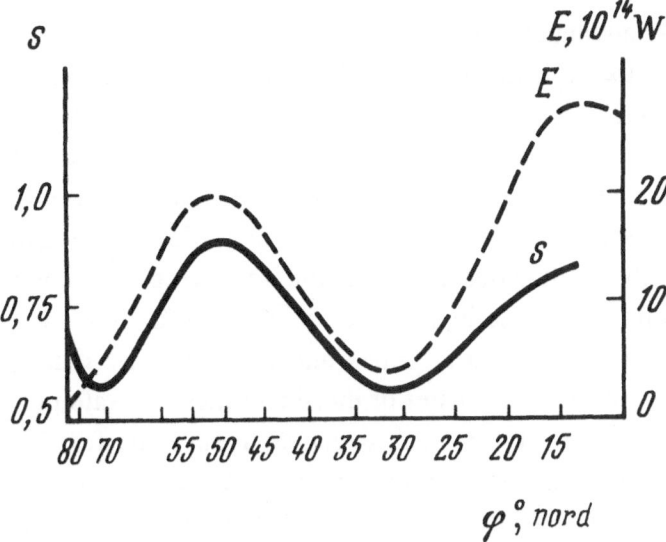

Fig. 2.3. The relative error of a one-year-ahead forecast of the average zonal air temperature (*s*) and the average total atmospheric meridional heat and potential energy flux (*E*). In the area between 75°N and 15°N where the temperature data are relatively good, the prediction error varies in the same way as for *E*. Assuming that the coincidence of the curves is not accidental, the predictability of statistical air temperature has a local character: as the meridional circulation of air decays, the forecasting errors for average zonal temperature diminish, and predictability rises. The one-step-ahead prediction errors are smallest in the area 30–40°N and at latitudes higher than 65°N. Probably, the increase in predictability within 30–40°N can be attributed to the two centers of atmospheric action located there, the Azores high and the Hawaiian high. The minimal relative prediction error at latitude 75°N almost corresponds to the location of the Greenland high.

2.7 Utilization of Forecasting Results

Americans are known to have a liking for counting money. It is perhaps for this reason that in 1972 Thompson estimated that the rational use ofmeteorological information in the USA effected a saving of $739 million annually. This comes from the use of advanced forecasting methods (about 60% of the overall saving) and from the benefits of improved efficiency resulting from economic decisions that are optimized on the basis of the forecasts. The major part of the saving, $567 million, is accounted for by the agricultural sector, and of the remainder, $112 million is saved by public utilities, $32 million by construction projects, $4 million by aviation, and so on. These figures are taken from a monograph by Zhukovskii [2.30], where the author systematically develops the methodology of using meteorological information to best advantage for solving various economic problems. We present a simple example that has recently excited public concern.

Suppose someone has predicted that next year the level of the Caspian Sea will fall with probability $p = 0.9$ or rise with probability $q = 0.1$. Must those authorized to take strategic decisions make the appropriate arrangements for dealing with a falling sea level? It is only natural to expect the authorities to make the following elementary calculations. Assume that a rise in the level of the Caspian Sea would submerge onshore oil production facilities and thereby cause damage valued at C roubles, whereas a transfer of those facilities to a safe place costs c roubles (obviously, $C \gg c$). A decision to save the facilities (decision 1) incurs expected costs $S_1 = cp + cq = c$, whereas a decision to leave everything as it is (decision 2), taking the prediction on trust, incurs expected costs $S_2 = 0p + Cq = Cq$. Thus, $S_2/S_1 = qC/c$. Now, if $C/c = 100$, S_2 is ten times greater than S_1, and decision 1 – the emergency evacuation – is clearly preferable, despite the small probability of submersion. Regrettably, all this happened in quite the reverse way in real life.

We can also see from this example that in evaluating a meteorological forecast we run up against two relatively independent problems [2.30]. The first one is assessing forecasting accuracy, and is of interest to those engaged in developing methods of prediction and to weather forecasters using these methods. The objective pursued when solving this problem is assessment of weather forecasting skills as such, irrespective of possible economic advantages. The other problem, more important to the user, is to evaluate the economic impact of the forecast so as to decide whether to use it or not, and if it is used, to estimate prospective benefits. Measures to evaluate both forecasting accuracy and economic impact are considered in great detail by Zhukovskii [2.30].

The above example leads us to the following scheme for making decisions based on forecasting results. Assume a forecasting result to be a set of probabilities $p_1, \ldots, p_n$ that one of the weather states $F_1, \ldots, F_n$ will be realized. This scheme easily incorporates observational data on continuously distributed values, in which case F_i is assumed to be an event falling within the i-th interval. In fact, it is precisely such continuous distributions, rather than discrete quantities, that interest users. Now suppose that a user can make a decision within a set of possible actions or measures $d_1, \ldots, d_m$, and that for each pair d_i, F_j an amount of damage $U(d_i, F_j)$ is predicted, where the damage is inflicted by weather event F_j after the decision to take measures d_i has been made. Of course, it may be not just damage but also profit, in which case we take U to be negative. Various terms are used to denote the function U in various fields of knowledge, for example, the utility function in econometrics and sociology, the gain function in game theory, and the loss function in decision theory. Let us take the last term. It is necessary to know the loss function to calculate an optimal strategy for using information. Expected losses through action d_k are

$$S_k = \sum_{l=1}^{n} p_l U(d_k, F_l),$$

so it is only reasonable for the user to choose action d_k such that it causes minimal losses S_k. In the above example, d_1 and d_2 are decisions 1 and 2, respectively; F_1 and F_2 are respectively rising and falling water levels in the Caspian Sea; $U(d_1, F_1) = c$, $U(d_2, F_1) = C$, $U(d_1, F_2) = c$, and $U(d_1, F_1) = c$.

A more flexible strategy is to prescribe probabilities $q_1, \ldots, q_m$, where $q_1 + \ldots + q_m = 1$, for taking one or another decision rather than to choose a particular action from the set $(d_1, \ldots, d_m)$. This is what we call a randomized decision-making strategy. It may be applied practically by altering actions $d_1, \ldots, d_m$ with frequencies $q_1, \ldots, q_m$, respectively. Sometimes randomization can be used in an isolated experiment. For instance, if action d_k denotes the process of manuring using a k-type fertilizer, the manured area is proportional to q_k. When a randomized strategy is used, the average losses are a function of the distribution $\mathbf{q} = (q_1, \ldots, q_m)$:

$$S(\mathbf{q}) = \sum_j \sum_l p_l q_j U(d_j, F_l) \tag{2.11}$$

The search for an optimal strategy is now reduced to a search for the vector $\mathbf{q}$ whose components minimize the value of the function (2.11). It is also very helpful to assume that decisions may be related to different weather conditions as dependent events. Then strategy becomes a set of conditional probabilities $p_{lj} = P(d_j | F_l)$ that decision d_j will be taken, given that weather conditions are F_l. This approach was originally suggested by Monin [2.32] and extended by Bagrov and Gruza together with Gelfand and Yaglom, the authors of [2.29].

The body of mathematics needed to tackle problems concerning optimal choice of strategy under uncertain conditions is now well developed, and it seems that in the search for more effective ways of using hydrometeorological information, major efforts should be directed toward the determination (by computation or estimation) of the loss function $U(d,F)$. As far as we know, such studies have not yet been initiated. There are numerous problems, such as that of giving numerical form to losses arising from human activities and the reaction of nature to such activities. Not all of these can be estimated in monetary terms. Whereas the profit from improved irrigation stemming from increased overland runoff can be calculated in roubles, the possible loss of certain species of fish (even those without value from a utilitarian point of view) in consequence of the increased runoff is hard to estimate.

This problem can probably be solved by the methods of non-numerical object statistics that are now being intensively developed [2.31]. In this approach, instead of using absolute quantitative characteristics of objects (or actions), one applies relative qualitative estimates, such as 'better' or 'preferable', and the loss function is optimized using the qualitative information.

2.8 Conclusion

It is hardly possible to give a complete view of the weather and climate predictability problem, in its current state, within the framework of this chapter. However, we hope we have presented a comprehensive enough survey to provide an insight into the basic forecasting methods, numerical characteristics of predictability, and factors limiting long-term forecasting.

We are grateful to Professor Yu. A. Kravtsov and Dr. V. E. Privalskii for valuable conversations that helped us improve the manuscript.

References

2.1   V.I. Tatarskii: On the criteria of chaos degree. Sov. Phys.-Uspekhi 32(5), 450 (1989)
2.2   A.S. Monin, A.M. Yaglom: Statistical fluid mechanics. Cambridge, MA: MIT Press, vol. 1, 1971; vol. 2, 1975
2.3   D.M. Sonechkin: Stochasticity in general atmospheric circulation models. Leningrad: Hydrometeoizdat, 1984 (in Russian)

2.4   N. Wiener, P. Masani: The prediction theory of multivariate stochastic processes. Acta Math. 98, 111–150 (1957)

2.5   A.S. Monin: Theoretical foundations of geophysical hydrodynamics. Leningrad: Hydrometeoizdat, 1988 (in Russian)

2.6   A.S. Monin: Weather forecasting as a problem in physics. Moscow: Nauka, 1969 (in Russian)

2.7   Yu.A. Kravtsov: Randomness, determinateness, predictability. Sov. Phys.-Uspekhi 32(5), 434–449 (1989)

2.8   I.A. Ibragimov, Yu.A. Rosanov: Gaussian random processes. Moscow: Nauka, 1970 (in Russian)

2.9   A.M. Yaglom: Extrapolation, interpolation and filtering of random processes with rational spectral density. Trans. Moscow Math. Soc. 237–238 (1955) (in Russian)

2.10 L.I. Piterbarg: Predicting one of random field classes. Probability theory and its application 1, 176–182 (1983)

2.11 J.E. Box, G.M. Jenkins: Time series analysis forecasting and control. San Francisco: Holden Day, 1970

2.12 L.I. Piterbarg: Dynamics and prediction of large-scale sea-surface temperature anomalies. Leningrad: Hydrometeoizdat, 1989 (in Russian)

2.13 K. Hasselmann: Stochastic climate models, part 1 – theory. Tellus 17(5), 473–485 (1965)

2.14 I.L. Vulis, A.S. Monin: Spectra of meteorological field long period variations. Proc. Acad. Sci. USSR, Doklady 197(2), 328–333 (1971) (in Russian)

2.15 E.N. Lorenz: A study of the predictability of a 28-variable atmospheric model. Tellus 17(5), 321–333 (1965)

2.16 V.E. Privalskii: Evaluating spectral density of large-scale processes. Izv. Acad. Sci. USSR, atmospheric and oceanic physics 12(9), 979–988 (1976) (in Russian)

2.17 V.E. Privalskii: Statistical predictability of large-scale hydrometeorological processes. Izv. Acad. Sci. USSR, atmospheric and oceanic physics 13(4), 371–377 (1977) (in Russian)

2.18 V.E. Privalskii: Climatic variability (stochastic models, predictability and spectra). Moscow: Nauka, 1985 (in Russian)

2.19 E. Parsen, J. Newton: Forecasting and time series model types of 111 economic time series methods. In: The forecasting accuracy of major time series methods, pp. 267–288. New York: Wiley, 1984

2.20 R.E. Davis: Predictability of sea surface temperature and sea level pressure anomalies over the North Pacific Ocean. J. Phys. Oceanogr. 6(3), 249–266 (1976)

2.21 A.M. Yaglom: Statistical forecasting. In: A.N. Kolmogorov (ed.): Transactions on the theory of probability and mathematical statistics, pp. 523–526. Moscow: Nauka, 1986 (in Russian)

2.22 E.N. Lorenz: Prospects for statistical weather forecasting. Statistical forecasting project, final report, p. 388. Cambridge, MA, 1959

2.23 A.N. Kolmogorov: On the matter of suitableness of statistically derived prediction formulae. In: A.N. Kolmogorov (ed.): Transactions on the theory of probability and mathematical statistics, pp. 161–167. Moscow: Nauka, 1986 (in Russian)

2.24 A.V. Meshcherskaya, L.V. Rukhovets, M.I. Yudin, N.I. Yakovleva: Natural components of meteorological fields. Leningrad: Hydrometeoizdat, 1970 (in Russian)

2.25 T.W. Anderson: Introduction to multivariate statistical analysis. New York: Wiley; London: Chapman and Hall, 1958

2.26 A.M. Obukhov. Izv. Acad. Sci. USSR, mathematics and natural science 3, 339–370 (1938) (in Russian)

2.27 H. Hotelling: Relation between two sets of variables. Biometrika 28, 321–377 (1938)

2.28 A.M. Obukhov, M.I. Fortus, A.M. Yaglom: The method of canonical correlations as applied to the analysis of random processes, and its utilization in meteorology. In: Abstracts of the first all world congress of the Bernoulli Society for mathematical statistics and probability theory 1, ser. 1–19. Moscow: Nauka, 1986 (in Russian)

2.29 I.M. Gelfand, A.M. Yaglom: Computing information about a random function that is contained in another one. Sov. Phys.-Uspekhi 12(1), 3–52 (1957) (in Russian)

2.30 E.E. Zhukovskii: Meteorological information and economic decisions. Leningrad: Hydrometeoizdat, 1981 (in Russian)

2.31 A.I. Orlov: Statistics of nonnumerical objects. Statistics, probability, economy, pp. 99–114. Moscow: Nauka, 1985 (in Russian)

2.32 A.S. Monin: On the use of unreliable forecasts. Izv. Acad. Sci. USSR, ser. geophysics 2, 218–228 (1962) (in Russian)

3. How an Active Autowave Medium Can Be Used to Predict the Future

G. R. Ivanitskii

Neural networks resemble autowave media composed of a multitude of oscillators. Their oscillation frequency varies depending on the intensity of a drive signal. Such networks are capable of predicting the future values of signal parameters based on past values.

3.1 Prediction

The bacterium has been ingested by a phage; the moth has been burnt in the flame of a candle; the hare has fallen prey to a lynx; the yacht's crew has perished in a storm; the inhabitants of a settlement have been buried under the ashes of an awakened volcano; the nation has become extinct because of an environmental disaster it engendered. These and other tragedies can all be attributed to one thing – erroneous estimation and prediction of the development of a situation. It is obvious that foresight is a factor in survival.

Irrespective of the means used to make predictions, either a computer or a biological system such as our brain, prediction is a physical process. The same kinds of concept apply to prediction as to other physical processes. Prediction, as the possibility of saying in advance that some event or series of events is going to happen, is based on the analysis and generalization of previous experience. The volume of information about the past varies from case to case. The overwhelming majority of events undergo periodic repetition (as will be shown below); however, not only are there synchronized cyclic processes in nature, but there are also recurrent manifestations of their desynchronization. In such events a process assumes a chaotic, random character. Chaotic behavior is usually unpredictable, though in itself the chaos may include a latent orderliness [3.1].

Predictions that may look unexpected at first sight are actually based on our past experience, the memory of which is either stored in the material history of our culture, folklore, and historical records, or comes to us via cosmic processes. A good deal of past experience is preserved genetically in living beings. It has been estimated that about one-third of the information man needs for survival is stored genetically, while two-thirds of it results from education and upbringing. These data were obtained in studies based on psychological testing of identical twins who were separated when young and raised by adoptive parents. In animals the role of the genetic component is greater than in human beings.

Our aim here is to show how active autowave media enable prediction of future events for a certain class of natural phenomena of a cyclic character. In particular, we shall consider a possible prognostic system that acts in much the same way as the human brain.

3.2 Active Autowave Media

The entire variety of media can be divided into two large groups representing conservative and active media. The former have no inner energy store, while the latter contain a diffuse potential energy store that may be accumulated in various ways. It is not always easy to draw a line between the two groups. Nevertheless, such a division is justifiable, as the reader will see later.

The concept of autowaves is a general one that allows systematization of both experimental facts and theoretical ideas about some nonlinear active-medium processes that are encountered in biology, chemistry, and physics [3.2–4]. The simplest autowaves are combustion waves, such as the flame front of a forest fire. Their propagation accompanies such processes as transmission of information in living beings, cardiac muscle contraction, primary morphogenesis stages in some protomorphic organisms, catalyst activation in the chemical industry, and so on.

It is important to note that autowaves occur only in active media and that their properties differ significantly from those of waves in conservative media, for example electromagnetic or mechanical waves.

The onset of oscillation in conservative media requires an initial disturbance to trigger wave propagation. The law of energy conservation is valid for such waves. When a stone is thrown into a stillwater lake, a part of the stone's kinetic energy is converted into the energy that produces circular waves on the lake surface. The laws of wave interactions for low-amplitude sinusoidal waves occurring in conservative media are simple. Because such waves easily penetrate each other, their interactions can be reduced to algebraic summation of the oscillations at each point in the medium (the superposition principle). This explains in particular the occurrence of the classic Moiré interference pattern formed by oscillating and calm regions of the medium (where wave amplitudes are added and subtracted, respectively; see Fig. 3.1a). Two other characteristic features displayed by common waves in conservative media, namely reflection from obstacles and interfaces and diffraction (bending around obstacles), can be also explained in terms of the fundamental superposition principle.

Although the energy of the initiating disturbance is preserved in a conservative medium, long-range signal transmission is beset with difficulties, since energy density in two- and three-dimensional media diminishes with distance from the source, while the form of the signal is distorted due to dispersion caused by the different velocities of the spectral components of the signal.

Unlike common waves, however, autowaves propagate in media with a dispersed energy store, that is, active media. It can be seen from Table 3.1 that the only feature shared by the two kinds of medium is that they support diffraction. The commonest active-medium example is operation of a safety fuse, in which the powder filling represents the energy store, while the autowave is a combustion wave travelling along the fuse. As the wave propagates, the fuse material converts from a high-energy stable state (unburnt powder) into a low-energy state (ash and gases remaining after the combustion is over) with part of the released energy dissipated and the rest used to initiate combustion of the yet unburnt length of the fuse.

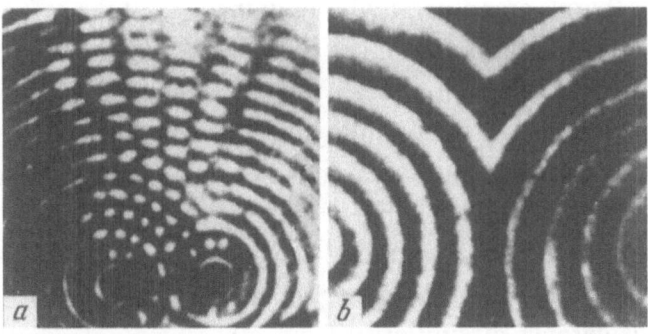

Fig. 3.1a,b. Interaction of waves coming from two sources: (a) wave interference on a water surface, (b) absence of interference for active medium autowaves (autowaves do not interpenetrate when they collide but annihilate)

Table 3.1. Properties of waves in conservative media as compared with those of autowaves in active media

Property	Waves[a]	Autowaves[a]
Inflow of energy	−	+
Preservation of amplitude and waveform	−	+
Reflection from obstacles	+	−
Annihilation	−	+
Interference	+	−
Diffraction	+	+

[a] Presence or absence of a property are indicated by + or −, respectively.

This case explains the following general definition. Autowaves are inherent in observable structures (in our particular example, the flame) which are maintained by processes of local release of stored energy used to initiate similar processes in neighboring areas. Such observable structures are called dissipative, since they originate from energy dissipation.

Long-range propagation of an autowave resembles a relay race in which a signal is newly reproduced at each point of the medium. The energy in the medium is not preserved but is consumed to maintain the autowaves, which explains the first two properties listed in Table 3.1. The energy of the autowaves is replenished from the medium. Collision of two autowaves leads to their annihilation (Fig. 3.1b). One can easily deduce this from the fact that following the running autowave front, bringing about the transition of the medium from the high-energy to the low-energy state, is a 'burnt' area (literally burnt in the case of the safety fuse) where a similar transition is no longer possible. As a result, two colliding autowaves destroy each other. We shall encounter this important feature again later. The impossibility of autowave interference and reflection from obstacles and interfaces can be explained in a similar way. However, autowaves are fully capable of bending around obstacles, or diffracting. As in the case of common waves in a conservative medium, diffraction here is explained by Huygens' principle (Fig. 3.2).

Due to their considerable dissimilarity from other types of wave, autowaves are set apart as a distinct class.

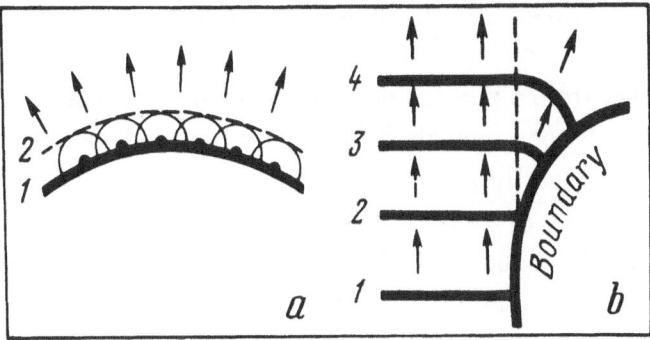

Fig. 3.2a,b. Autowave diffraction. (a) Huygens' construction: each point of the medium approached at a given instant of time by an autowave front becomes a source of circular autowaves, (b) an autowave bending around the active medium boundary (consecutive front positions are numbered)

3.3 Autowave Propagation in Energy-Restoring Active Media

The most interesting autowaves are those travelling in an active medium with a replenishing energy store. We shall use such a medium later to construct a prognostic system. In the course of slow recovery, a reverse process of medium transition from the low-energy state (caused by the autowave) to the initial high-energy state takes place. Such media may contain local self-maintained autowave sources of diverse types.

Let us consider three examples of energy-restoring active media: a burner with wicks that slowly draw up oil, a chemical active medium, and a biological excitable tissue. Regularities of autowave propagation and interaction are independent of the physical form an active medium takes.

A burner-type active medium. A special burner was designed in 1979 by Morozov and the author to demonstrate the simplest mechanism of autowave initiation and propagation. A number of holes were drilled close to one another in a metallic sheet, and asbestos strips were inserted through the holes. The top ends of the strips were connected and the bottom ends were immersed in a bath of thick oil. Asbestos is not imflammable by itself, so when soaked with oil it can serve as a wick. The combustion rate of the oil-soaked asbestos wick exceeds the rate at which oil is drawn up. Consequently, the flame at the top of the wick will be extinguished after a while. Oil will then seep up the wick by diffusion, allowing the wick to be ignited again. We can thus define three stages in the operation of the wick: the combustion stage, the pause (the refractory period) while oil is absorbed, and the stage of readiness for renewed ignition.

Ignition passes over from a burning wick in such a burner to an unlit wick. The first wick will soon quench as its oil runs out, but the flame front will already be travelling further along the burner. Thus a technically simple model of a recovering active medium can be constructed. Each element of such a medium (each wick) can be ignited again and again, unlike the safety fuse. It is important to note that repeated ignition is possible using not only an external source but also using a flame delivered via the medium. This simply requires a closed circuit of flame-transferring wicks. In the case of a planar array of wicks, under certain conditions a rotating fire vortex or circular wave propagation will be observed (Fig. 3.3).

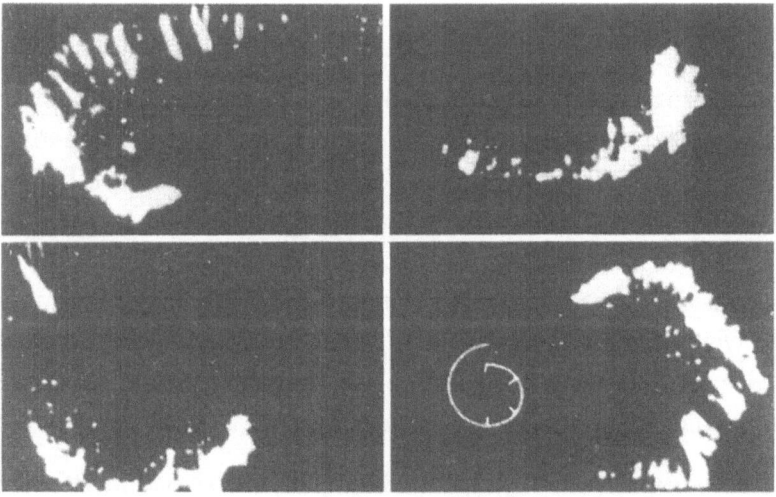

Fig. 3.3. Rotation of the spiral front of the burner's flame: shown schematically on the left of the last frame are phases of the front's rotation (Morozov and Ivanitskii)

An active medium based on a chemical change. A chemical active medium was first prepared in the Institute of Biophysics, USSR Academy of Sciences, in 1970 by Zhabotinskii and Zaikin. It consists of a thin liquid layer in which the Belousov oxidation reaction (otherwise known as the Belousov–Zhabotinskii reaction) takes place. The medium exhibits a cyclic (oscillatory) behavior pattern. Unlike the majority of oxidizing processes, which go on until one of the substrates (the oxidizer or the reducer) is exhausted, the reaction produces an agent inhibiting its course for some time following depletion of just a small proportion of the reagents. The composition of the reaction mixture described by Belousov in the mid-1950s is as follows: 2.00 g citric acid, 0.16 g cerium sulfate, 0.20 g potassium bromate, 2.0 ml sulfuric acid (1:3), and 10.0 ml water to fill the volume. Cerium, which is a variable-valence metal, plays the role of a pendulum, since it may participate as both oxidizer and reducer.

The general outline of the reaction is as follows. The bromate oxidizes Ce^{3+} ions to Ce^{4+} and transforms into a hypobromite, which in turn oxidizes the citric acid, while bromine reduces Ce^{4+} to Ce^{3+}, yielding a bromine ion capable of inhibiting the reaction between Ce ions and the bromate. An alternating variation in the extent of cerium oxidation shows up as periodic changes in the color of the solution. Other substrates can also be used to initiate a periodic reaction – much than 50 such reactions are known so far. A more comprehensive description of a similar reaction was later given by Neues and Field (USA) and Keros (Hungary).

If a thin layer of the reaction mixture given above is poured into a flat vessel, then, because of impeded dispersion, each small liquid volume may be regarded as an element of a recoverable active medium that can be traversed by autowaves as many times as the substrate stock allows (Fig. 3.4). In the Belousov–Zhabotinskii reaction, about 1% of the substrate in a volume element oxidizes each time an oxidization wave traverses it, so waves can pass over it up to 100 times. The mechanism of oxidation-wave propagation is basically the same as with the combustion waves, since combustion is a particular case of oxidation where excited (burning) medium elements excite (ignite) their neighbors.

An active medium in the form of an excitable biological tissue. Systematic research into biological active media was initiated in the Institute of Biophysics in 1966 by Krinsky and his coworkers.

The commonest example of such a medium is nervous tissue. An impulse travelling along a nerve fiber is actually an autowave in the form of an electrochemical wave transient between the rest state when the potential difference across the tissue membrane is high (about −0.08 V) and the excited state when the potential difference is low (less than +0.04 V). Propagation of a nerve impulse is accompanied by the release of energy originally stored in the form of unbalanced concentrations of sodium and potassium ions on both sides of the membrane (Fig. 3.5).

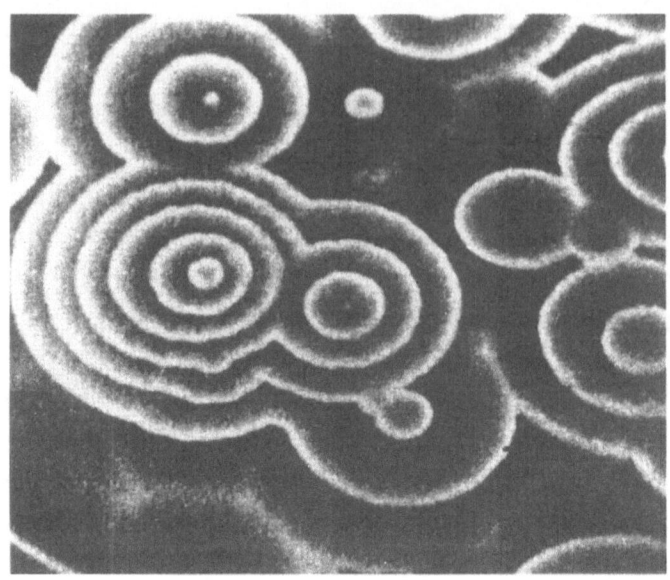

Fig. 3.4. Autowaves in a chemical active medium (Zhabotinskii and Zaikin)

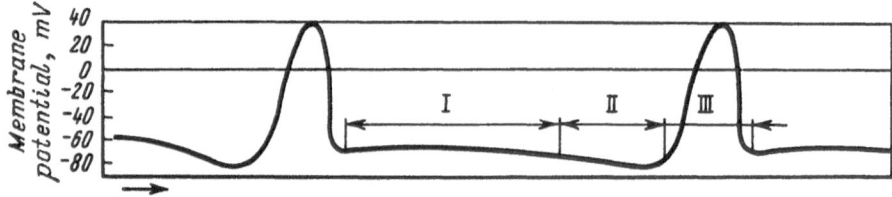

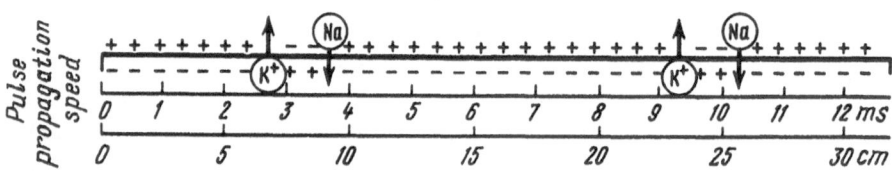

Fig. 3.5. Propagation of an excitation pulse (a wavefront) along a nerve fiber. *Top:* pulse parameters. Roman numerals designate periods in which the nervous tissue goes through different stages: I – rest period, II – recovery period (refractoriness), III – excitation period. *Bottom:* Schematic representation of depolarization of nervous tissue at the instant of passage of the pulse

Living organisms contain not only one-dimensional active media in the form of nervous tissue but also two- and three-dimensional active media consisting of excitable cells functioning in much the same way as nervous tissue.

Examples of such media are brain and spinal marrow, nonstriated-muscle walls of gut, uterus, and urinary bladder, and cardiac muscle. Autowaves travelling through them are of the same physical nature as nervous impulses, differing only in duration and speed. However, the roles they play in providing vital functions are different. Whereas an impulse running along nervous tissue enables the transmission of information, an excitation wave in a cardiac muscle triggers a series of biochemical processes that initiate muscle contraction, the mode of which depends on variations in the propagation of the excitation wave.

The mechanisms underlying the operation of all active media (physical, chemical, and biological) are similar and can be described mathematically by the same expressions. These expressions have the form of parabolic-type differential equations:

$$\frac{\partial U}{\partial t} = D_U \Delta U + f(U,V),$$

$$\frac{\partial V}{\partial t} = D_V \Delta V + \varphi(U,V),$$

(3.1)

where U is (a) combustion energy in the oil-burner model, (b) the active radical concentration in the chemical reaction model, or (c) the potential difference across the cell membrane in the biological tissue model; V is (a) the required amount of combustible oil in the burner model, (b) concentration of oxidized catalyst in the chemical reaction model, or (c) cell membrane conductivity in the biological tissue model; D_U and D_V are the coefficients of diffusion of excitation and recovery (refractory) waves; t is time;

$$\Delta = \frac{\partial^2}{\partial x^2} + \frac{\partial^2}{\partial y^2} + \frac{\partial^2}{\partial z^2}$$

is the Laplace operator; and $f(U,V)$ and $\varphi(U,V)$ are the functions establishing the relationship between parameters U and V. The pattern of this relationship is governed by behavior of the system at a point. We shall resume discussion of this relationship in Sect. 3.4.

In all these cases, the active medium can be regarded as a two-level structure that may exist in either a high-energy or a low-energy state. As an autowave propagates, medium elements on its front jump downward from the high-energy state. The energy released is used, as stated earlier, to launch similar jumps in the regions of the medium immediately in front of the autowave.

Elements of an active medium that are unable to restore their energy remain in the low-energy state after the downward transition and thus block the repeated propagation of an autowave (for example, in safety fuse combustion and phase-transition waves). In energy-restoring active media, autowaves may in principle travel any number of times, since each element of such a medium reverts to the high-energy state following the slow energy uptake which constitutes the energy restoration process.

Recall once more that for a burner with slow-absorption wicks, the high-energy state corresponds to wicks saturated with oil, and the low-energy state to dry wicks that have not yet soaked up oil. For excitable cells, the high-energy state corresponds to a large potential difference (−0.08 V) between the inner and the outer sides of the cell membrane, and the low-energy state to smaller values of this potential difference (+0.04 V). It is not until the transition to the high-energy state is completed that a medium element normally becomes excitable, and the time needed for this transition is called the refractory period (a term borrowed from physiology).

3.4 Dynamics of Autowave Interaction

In active media, one can find autowaves with standard configurations that may be grouped by similarity of appearance and formation mechanism.

At least four autowave types can be distinguished: concentric waves, otherwise known as pacemakers; spiral waves, or reverberators; cellular waves, e.g., Bienar cells; and spatial vortices (Fig. 3.6).

Autowave interaction dynamics are characterized by three main features. First, there are two interchangeable modes of autowave source activity, the standby (one-shot) mode and the auto-oscillatory (or self-oscillatory) mode. Second, autowave sources compete for space in which to emit waves. Third, the frequency of self-oscillation may be put under control.

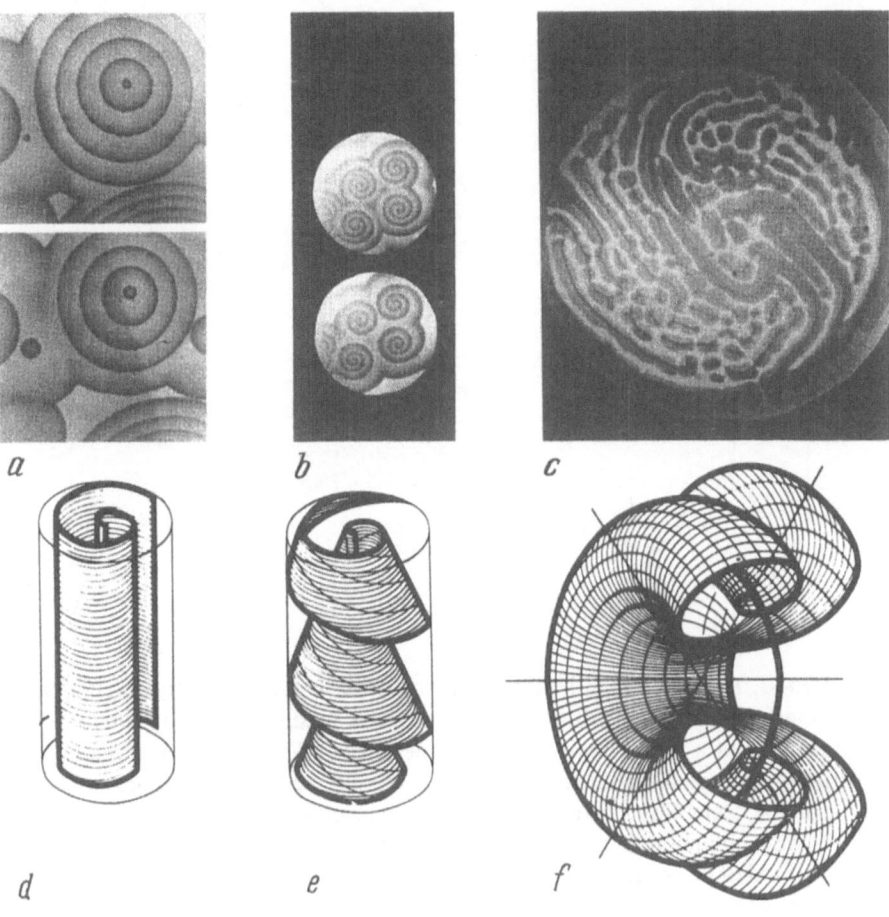

Fig. 3.6a–f. Different types of autowaves occurring in an active medium in which energy is restored: (**a**) concentric autowaves, (**b**) spiral autowaves, (**c**) cellular autowaves, (**d**) ordinary spatial autowaves (vortices), (**e**) twisted spatial autowaves, (**f**) vortical spatial rings

Equation (3.1) describes the spatial behavior of active media. The relevant concentration system given by

$$\dot{U} = f(U,V)$$
$$\dot{V} = \varphi(U,V)$$

$$(3.2)$$

is actually a nonlinear relaxation system with an N-shaped characteristic (Fig. 3.7), called the generalized Van-der-Pol model, which may serve both as a standby single-pulse oscillator (mode 1) and as a concentric-wave auto-oscillator (mode 2, Fig. 3.7b).

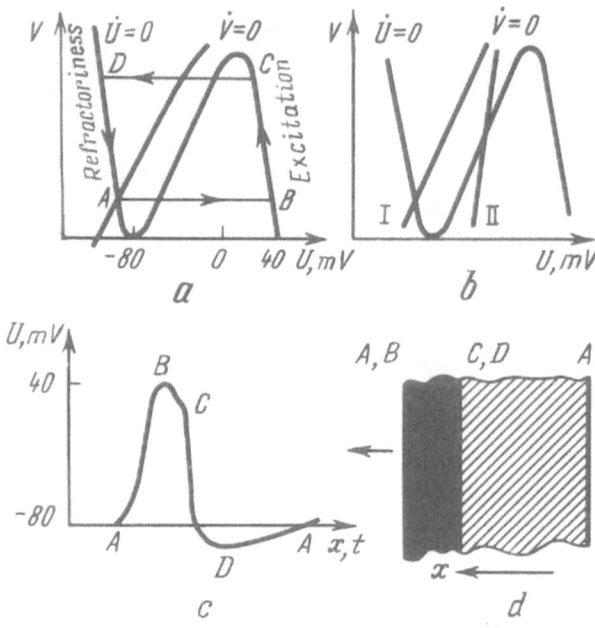

Fig. 3.7a–d. A phase plane indicating the relationship between the excitation and inhibition (refractoriness) components of an active medium given by equation (3.2). (**a**) trajectory of a representative point in the phase plane of a concentrated system generating an excitation autowave in nervous tissue (see Fig. 3.5); portions BC and CDA of the trajectory correspond to the states of excitation and refractoriness, respectively, and point A corresponds to the state of rest, (**b**) a phase plane indicating the two modes of operation: I – a single-pulse oscillator, II – an auto-oscillator; (**c**) profile of potential change in nervous tissue corresponding to the representative point trajectory in the phase plane; (**d**) diagram of an autowave fragment corresponding the the potential change profile in c (the excited region is shaded and the refractory region is cross-hatched)

The system can be switched from the one-shot mode into the auto-oscillatory mode by changing the relative positions of the $\dot{U}$ and $\dot{V}$ functions on the phase plane. This transition can be achieved either by speeding up energy delivery into the excited system or by accelerating energy consumption in the course of excitation, i.e., by increasing the gain of the excitation circuit or the reduction of the gain in the inhibition circuit.

Let us assume that two auto-oscillators produce concentric autowaves propagating in an active medium. The frequencies at which they emit the waves are ω_1 and ω_2, respectively. As mentioned above, collision of autowaves results in their mutual annihilation. The auto-oscillator with the higher emission rate will capture step by step (with each emission cycle) the space controlled by the oscillator with the lower frequency and eventually absorb it. The time taken for this expansion is

$$t = \frac{l}{v}\left(\frac{\omega_1 + \omega_2}{\omega_1 - \omega_2}\right) \tag{3.3}$$

where l is the distance between the two auto-oscillators, v is the autowave propagation rate, which is constant for the given medium, and ω_1 and ω_2 are the circular autowave emission frequencies for the respective auto-oscillators. If $\omega_1 = \omega_2$, then time t becomes infinitely long, which means that two auto-oscillators with the same emission frequency cannot absorb one another. When ω_2 is close to zero, the time taken for the first auto-oscillator to capture the domain of the second is minimal.

Figure 3.8a shows a geometrical interpretation of the interaction between two autowaves emitted simultaneously by two different auto-oscillators operating at different frequencies. The autowaves can be conceived as circular conic surfaces. The higher the auto-oscillation frequency, the the larger the angle between the cone axis and its generator. Figure 3.8b illustrates the dynamics of the spatial expansion of the higher-frequency autowave over time. A similar process taking place in a chemical active excitable medium is shown in Fig. 3.9.

The operation frequency of an auto-oscillator in an active medium can be controlled, i.e., increased or reduced by varying the energy consumption rate in the excitation circuit or the rate of energy restoration via the inhibition circuit. In neural networks in nervous tissue, this process is implemented by the excitation or inhibition of synaptic terminals through secretion of corresponding mediators which influence ion channels in neurons. In the burner case discussed above, such control can be exercised by changing the viscosity of the oil, by adding a solvent periodically, or by changing the section of the wick or pinching it. However, from the standpoint of technical realization it is better to use media in which oscillator frequency can be controlled by electrical or light signals.

This idea may be implemented using, for example, a chemically active medium, formulated on the basis of the Belousov–Zhabotinskii reaction, to which a light-sensitive ruthenium-based catalyst has been

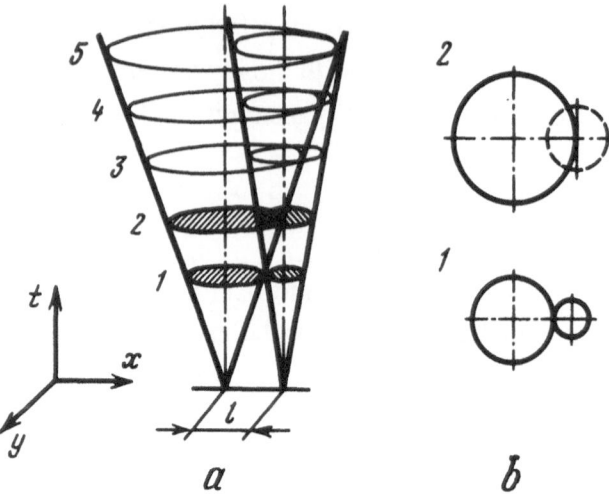

Fig. 3.8a,b. Geometrical interpretation of the interaction of autowaves emitted by two auto-oscillators with different frequencies. The vertical axis represents time. The horizontal axes are the plane coordinates. The higher the auto-oscillation frequency, the larger the angle between the cone axis and its generator. (**a**) interaction of autowaves from the two oscillators spaced distance l apart. (**b**) the oscillator with the higher frequency capturing the the domain of the oscillator with the lower frequency by means of the annihilation of colliding autowaves. Two cone cross-sections are given, the first showing the instant at which the autowaves meet, the second the instant at which the higher-frequency auto-oscillator suppresses the lower-frequency one. At the latter instant, the autowave from the higher-frequency oscillator overlaps the emitting center of the lower-frequency oscillator

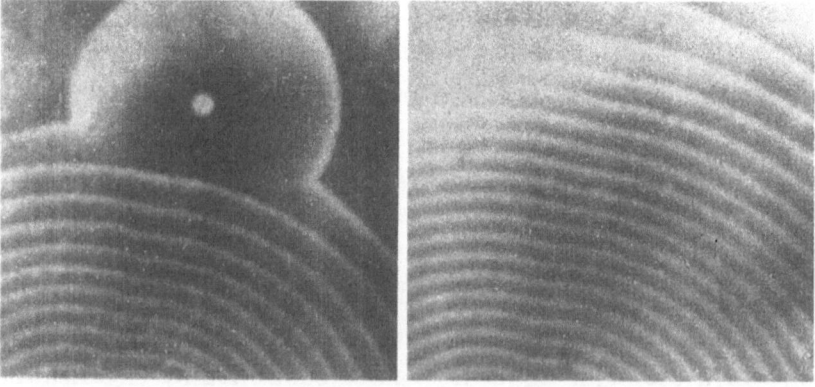

Fig. 3.9. Annihilation of the lower-frequency oscillator by autowaves from the higher-frequency oscillator during the Belousov–Zhabotinskii reaction

added, or using a set of photodiodes switched into the feedback circuit for the active elements of the medium. To form an active medium, one can employ liquid-crystal matrices, light-sensitive protein molecules, or special photochromatic glasses [3.5–3.8]. What is important in principle is that some active autowave media may contain matrices of auto-oscillators whose frequencies vary in proportion to the intensity of an external signal.

3.5 The External Medium Model and Its Fourier Image

We need a model of the behavior of the external medium to make predictions. If the model is based on the kind of postulates set out above, then we need to circumscribe it with some explanatory comments. However, this will confine our freedom of choice, since such circum-scription generally introduces arbitrariness that does not exist in nature.

It makes no sense to ask how many models of the external world can be invented. They are infinite in number. When various phenomena coexist, it is more pertinent to speak of their relationship and com-plementarity than of priorities.

We choose to restrict our attention to a cyclic model of the orga-nization of the external medium, which is simple enough to admit a variety of forms of behavior. Although the model enables predictions, it requires no stringent circumscription as to what concerns cause and effect. Periodically, 'cause' and 'effect' change places in the cycle, depending on when and where we start our observation. Indeed, there is a strong periodic component in many processes taking place inside and outside us. Even the so-called 'arrow of time' should be regarded as a process developing periodically, though at rather large periods.

When microbodies come into motion, separate cycles easily dis-cernible in microprocesses merge due to dephasing, and we describe the motion in terms of continuously increasing time. This actually underlies the determinism of the classical laws of mechanics. In such processes, long-term prediction is possible. But what is a long term? In comparison to what is it long? Let us consider some examples.

Modern Newtonian celestial mechanics, corrected using Einstein's relativistic law of gravitation, has a predictive power that is effective for millions of years. The scale of the evolution of the universe is billions of years. On such timescales, the influence of so-called minor parameters may be quite significant, and hence it is possible that the universe includes cyclic processes with periods much larger than, say, the time for which our planetary system has existed.

One should therefore presume that the term for which a prediction is made (the predictability horizon) is shorter than the stability period of the given system, i.e., the interval between two bifurcation points on the system's development trajectory. If our knowledge about the periodicity of this trajectory alteration is full enough, i.e., if plenty of information on the behavior of the system between bifurcations is available, then the prediction horizon can be expanded.

Consider the example of a smaller, yet still large time interval of 600 million years separating us from the earliest multicellular living organisms. Paleontological fossil remains suggest that periods of mass extinction took place within this time interval every 26–28 million years or so. For instance, it was 65 million years ago that the dinosaurs became extinct. However, long-term forecasting of the macroprocess here is difficult due to the scarcity of fossil remains available for investigation.

Obviously, there are shorter periods in the development of the external medium. The period of the revolution of the Earth around the Sun is 365.25 days. The revolution of the Moon around the Earth has a period of 29.53 days. The seven-day week characterizes the periodicity of atmospheric fluctuations correlated with ionospheric changes caused by the changing flux of charged particles emitted by the Sun – the so-called solar wind. The rotation of the Earth about its axis produces a daily (circadian) rhythm. Within the biosphere, one comes across a variety of rhythms governed by the life cycles of different populations, including individual species. There are generational cycles with periods of 25–27 years in cities that arise from the human reproductive cycle – an example of a social rhythm. The list of cyclic processes can be extended, but these examples seem sufficient to show that the world we live in can be seen in large part as an interaction of periodic processes. So why, one may ask, is it so difficult to make predictions within such a well-ordered structure?

3.6 Non-isochronism of Cyclic Processes

In the majority of cyclic systems, the amplitudes and frequencies of cycle-forming 'pendulums' are dependent on energy acquired from the outside by the system concerned. The variation of energy values is dependent in turn on the movement of pendulums belonging to other hierarchic levels. The feature of frequency dependence upon energy is called non-isochronism. The pattern of this dependence may vary according to the way the system is organized, e.g., frequency may increase stepwise with energy raised to the power of 3/2, 1, or 1/2.

Oscillation non-isochronism is characterized by frequency variation per unit of acquired energy variation. The greater this quantity, $d\omega/dE$, the less the inertia of the 'pendulums' and the sooner they come into and out of resonance (i.e., the shorter the span between bifurcation points), and eventually the more complicated prediction becomes within such a system.

In the system of pendulums, resonances may arise with pendular frequency ratios not only of 1:1 (oscillating in unison) but also of 1:2, 2:5, 1:3, 1:4, 1:5, 3:5, 3:7, etc. This fact explains how numerous satellites can revolve around a planet with different orbital periods. However, delivery of a large amount of energy into a system in which interpendular bonds are weak may cause migration of pendular frequencies over different resonance ratios. Such frequency-scale migration was familiar already to Poincaré, who remarked: 'the complexity of this motion pattern is so striking that I do not even try to describe it'. During migration, frequency trajectories cross each other in the phase space and cause the system to wander over resonances without bound, a phenomenon called Arnold diffusion after V. I. Arnold, who described its dynamics to a first approximation [3.9]. When this phenomenon occurs, the swings of different pendulums between resonances become phase-desynchronized.

Conversely, if there is no Arnold diffusion, the bonds are strong, the energy input is small, and the swings are synchronous. All other bond-energy parameter variations produce intermediate cases. Figure 3.10 illustrates computer simulation results of the dynamics of two non-isochronous pendulums with strong bonds between them. At small energy values, the swings are synchronous. Then, with rising frequency, the reciprocal swinging becomes dephased. Phase opposition occurs. With growing energy delivery, the pendulums start swinging independently with random phase alterations, brining about a chaotic state. A reverse development, following energy decrease, yields a hysteresis with the loop depending on the $d\omega/dE$ factor of the frequency–energy lag in a two-pendulum system.

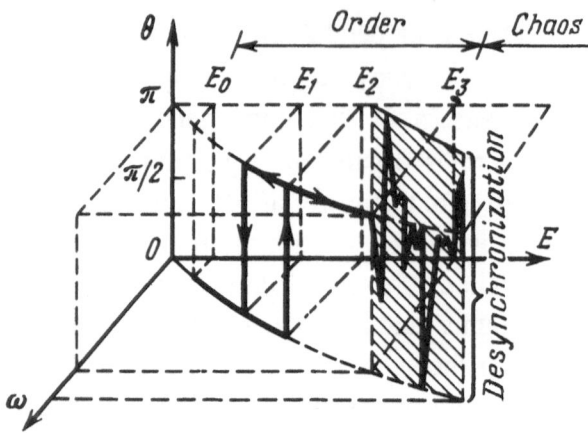

Fig. 3.10. Varying frequency ω and oscillation phase θ for two interacting pendulums as functions of input energy E (Ivanitskii and Morozov)

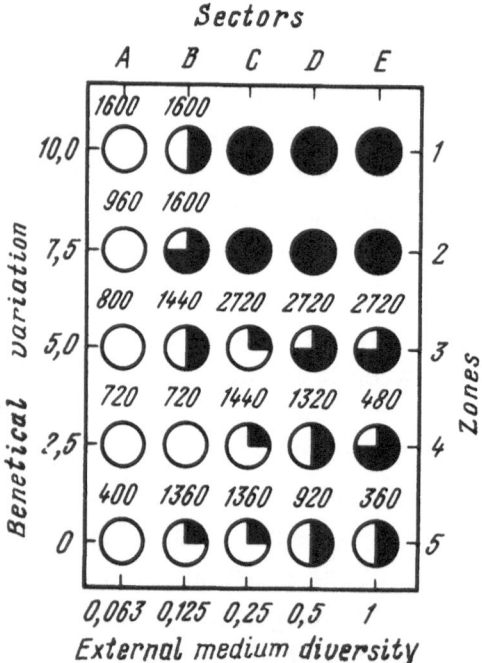

Fig. 3.11. Results of computer simulation of phase changes in a system of n pendulum oscillators. The pattern of pendulum interaction is governed both by their structure (genetic variation) and by externally introduced energy (external medium diversity). If the two parameters vary, one can observe at least five patterns of group behavior, ranging from complete phase coincidence to complete dephasing (Ivanitskii, Morozov, Rusanova)

Increasing the number of oscillators results in a larger variety of behavior patterns [3.10]. Figure 3.11 shows a diagram representing computer simulation results for a multi-oscillator case. A number n of non-isochronous oscillators with straggling parameters, i.e., with genetic variation (measured in relative units, from 0 to 10), are placed in a medium. The variation depends on the constructional parameters of the oscillators and on the spread of initial oscillation frequencies. The oscillators are connected via their common power source, which in turn acquires energy from another source of unbounded volume. The initial filling of the common power source (the external medium diversity) may vary over the range from 0.063 to 1 (in relative units). The behavior of such oscillators is peculiar. If the common power source is low, the non-isochronous oscillators become synchronized after a certain number of observation cycles (shown by numerals in the diagram) and their pulsation continues in unison (open circles, sector A, zones 1–5). With an ample power source and extensive genetic variation, there is no synchronization at all (closed circles, sectors C–E, zones 1–2). In the latter case, the system displays chaotic behavior. For other genetic-variation parameter values and external medium types, intermediate behavior patterns show up. One can observe oscillator synchronization with phase incursion (sector–zone indices B1, D4, E5). Synchronization invloving breakdown of oscillators into groups acting in unison is also possible. The groups themselves are phase-shifted (sector–zone indices B5, C3, C4, C5). Finally, there may be a partial synchronization with oscillators appearing in between phase-shifted groups oscillating in unison. Those are double-frequency 'conformist' oscillators, joining now to this, now to that group. They produce an impression of partial synchronization, resembling complete desynchronization that has taken the form of chaos (sector–zone indices B2, D3, E3, E4).

3.7 Harmonious Modulation and Modulation of Harmonics

Any physical process can be viewed as a space- and time-dependent function that can be disintegrated into a sum of other functions. For periodic processes, the most suitable form of series expansion is known to be an expansion in terms of harmonic components, i.e., expansion into a Fourier series. Such a series converges readily and is limited by the highest frequency of the periodic process. Direct and inverse transitions from the time domain into that of Fourier space frequencies are always possible.

Let us now formulate one of the prediction problems. In what way will a non-isochronous oscillator behave within a system of associated oscillators? Which of the frequencies in the variation of its behavior are significant and most representative of the 'oscillator–medium' system? What limits the accuracy of prediction?

We shall begin with the last question. For a one-dimensional case, a non-isochronous periodic process can be characterized as follows:

$$y(t) = a_1 \sin[\omega_1 t + \theta(t)], \tag{3.4}$$

where $\theta(t)$ is a phase change. If we know how function $\theta(t)$ varies, we can make a good forecast for any length of time. Assuming that the world we live in is a system of communicating vessels with different energy-retention capacities and correspondingly different sets of energy-exchange cycles, the behavior of the exterior medium can be described in the form:

$$y(t) = a_1 \sin[\omega_1 t + \theta_1(t)],$$
$$\theta_1(t) = a_2 \sin[\omega_2 t + \theta_2(t)],$$
$$\theta_2(t) = a_3 \sin[\omega_3 t + \theta_3(t)], \tag{3.5}$$
$$\cdots \cdots \cdots \cdots \cdots$$
$$\theta_n(t) = a_{n+1} \sin[\omega_{n+1} t + \theta_{n+1}(t)],$$

where $a_2, a_3, a_4, \ldots$ are the amplitudes of the signals modulating the fundamental frequency, or, otherwise expressed, the amplitudes of the frequency deviations. The choice of fundamental frequency is governed by the length of time for which the prediction is made.

Let us restrict our discussion to just two hierarchical levels. Figure 3.12 shows three examples of process behavior in the time domain with frequencies of a non-isochronous oscillator following the harmonics:

$$y(t) = a_1 \sin[\omega_1 t + a_2 \sin \omega_2 t]. \tag{3.6}$$

Figure 3.13 presents corresponding fragments of the Fourier images.

With ω_1 approaching ω_2 and $\varepsilon = a_2/\omega_1$ exceeding unity, the process approximates a chaotic one; in Fig. 3.12c, for example, $\varepsilon = 5$ and $\omega_2/\omega_1 = 3.45$. When frequencies ω_1 and ω_2 differ by several orders of

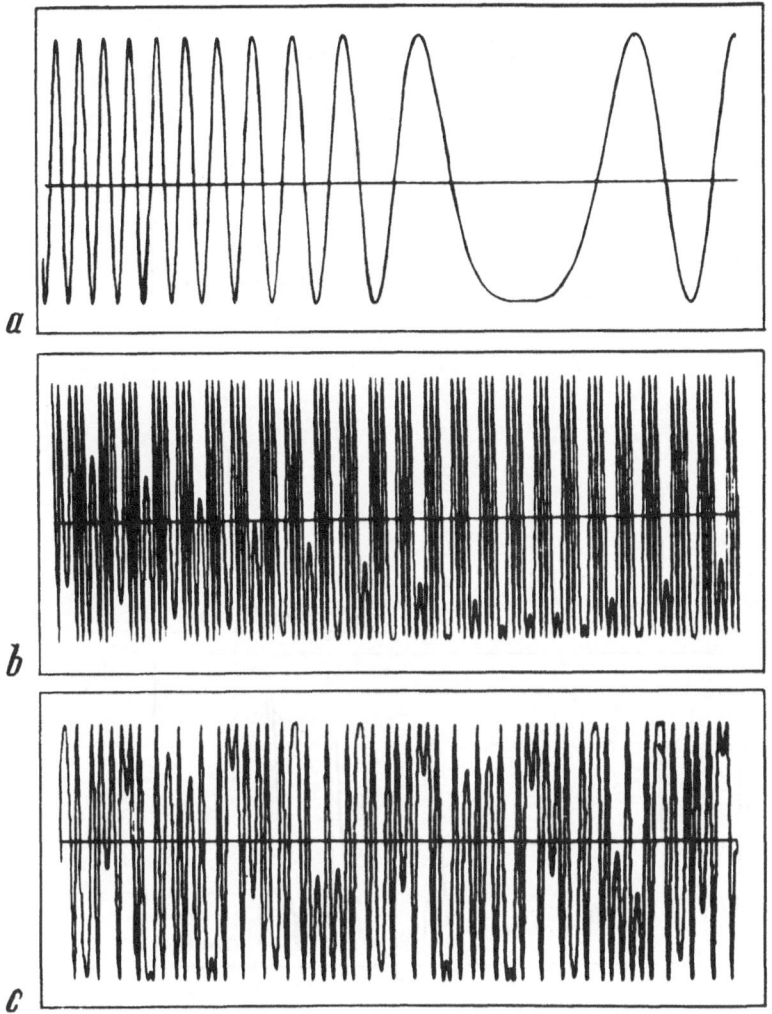

Fig. 3.12a–c. Evolution of phase-modulated sinusoidal process fragments with time. Phase modulation is governed by a sinusoidal law. The parameters of the process are:
(a) $\varepsilon = 50$, $\omega_2/\omega_1 = 0.034$
(b) $\varepsilon = 10$, $\omega_2/\omega_1 = 36\,000$
(c) $\varepsilon = 5$, $\omega_2/\omega_1 = 3.45$

magnitude, the process still varies periodically, even at high ε; in Fig. 3.12a, $\varepsilon = 50$ and $\omega_2/\omega_1 = 0.034$. A similar pattern is followed with the reverse ratio of frequencies; in Fig. 3.12b, $\varepsilon = 10$ and $\omega_2/\omega_1 = 36\,000$.

The conclusion is clear. The impression of chaos stems from the relative proximity of frequencies, both those of the process itself and those of the frequency-modulating power source. Given that ω_2 is either much more or much less than the fundamental frequency of the process,

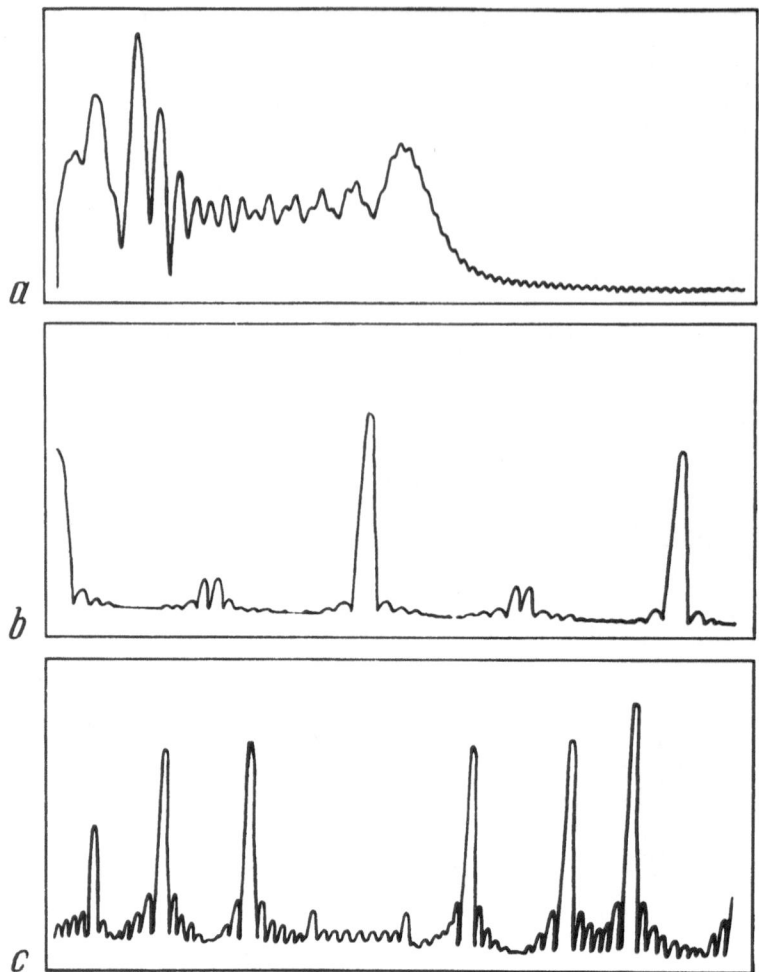

Fig. 3.13. Fragments of Fourier images in the frequency domain corresponding to the processes shown in Fig. 3.12

the superfast or superslow modulating components can be neglected in considering the process over a given time interval. It is important to note that the choice of a prediction term is given also by the structural parameters of the prognostic system itself. Living organisms, including man, are no exception.

In the analysis and executive systems in living organisms, high-speed phenomena which proceed one after another tend to merge together, and are perceived as a unity. For instance, two light flashes separated by 100 ms are discerned as one flash. The temporal resolution of sensory systems in living organisms is coupled by evolution with the reaction times of their motor systems and with the changeability of their active medium, subject to the survival constraints imposed by their ecological niche.

These sensory and analysis systems are nothing more than biochemical machines in which information is carried by flows of ions, which are relatively large and low-energy particles. Ions give rise to auto-oscillations in the brain's active medium. Therefore, the capability for spatiotemporal analysis of the active medium in the animal brain has upper bounds set by regions occupied by relatively large oscillators (the minimal spacing between ion channels in nervous cell membranes is 8–10 nm). Active media composed of low-energy information-carrying particles can respond only to combined variations of high-energy particles, such as photons, because at each space-time point in a medium analyzing an external process, a multitude of wave packets associated with various small high-energy particles overlap. This simple quantum-mechanical principle implies that the frequency-analysis capability of the human brain has an upper bound of approximately 10 Hz.

Such an assessment is supported by yet another consideration. The duration of a neuron excitation pulse is about 3 ms, and its refractory tail is almost twice as long as the pulse itself (Fig. 3.5). Hence the computing speed at each point of an active autowave medium composed of neurons is not more than 100 operations per second. A nervous pulse travels along an axon at a speed of the order of 20 ms^{-1}. The average path length between analysis and executive elements is about 1.5 m. To cover it takes about 75 ms. Therefore, the upper frequency is close to 10 Hz. It is worth noting in this connection that the dominant rhythm in our brain, the alpha rhythm, which shows up as early as 5–6 years of age and is probably representative of the fundamental frequency of the brain's active medium auto-oscillators, also falls within the range of 8–13 Hz. To sum up, we state that the predictive power of which the human brain is capable (unaided by any tools) has an upper frequency bound of 10 Hz.

The lower frequency limit of the brain depends on the organism's memory duration, i.e., its lifespan. Note that we do not touch upon social and technical means of expanding this range, such as historical documents, ancient relics, archives, and so on. Human memory and acquired knowledge have a number of peculiar features. For example, the war of 1941–1945 seems as remote to contemporary youths as the war of 1812. Historical events that happened before one's birth tend to level off chronologically in one's consciousness. For veterans, on the other hand, the events of World War Two form vivid and emotionally colored reminiscences.

3.8 The Fourier Image Cleared by the Active Autowave Medium

We now come to the techique that executes the predictive function of the active autowave medium. A Fourier image of an external process is projected during some time interval onto a two-dimensional autowave medium composed of an auto-oscillator matrix (Fig. 3.14). The medium has to synthesize the development of this process with the highest possible authenticity. Auto-oscillation frequencies vary with the harmonic intensity of the Fourier image. Thus geometric projections of auto-oscillator emissions (Fig. 3.8) indicate the distribution of harmonic intensity in the Fourier space. As mentioned in Sect. 3.4, the dynamics of auto-oscillator interaction in active media is such that lower-frequency oscillators (low-intensity Fourier image harmonics) become suppressed in the vicinity of higher-frequency ones (high-intensity Fourier image harmonics) after some time interval.

The suppression area S_i in the i-th zone of the Fourier space grows over time in agreement with

$$S_i = \pi \left\{ \frac{1}{n} \sum_{k=1}^{n} v^2 \left[t_{i-k} + t \left(\frac{\omega_i - \omega_k}{\omega_i + \omega_k} \right) \right]^2 \right\}, \qquad (3.7)$$

where t is the running time of the observation, t_{i-k} is the instant at which the i-th oscillator autowave encounters the i-th wave of the k-th oscillator, ω_i and ω_k are the respective frequencies of the autowaves emitted by the i-th and k-th oscillators, v is the speed of autowave propagation over the medium surface, and n is the number of oscillators whose autowaves are in contact with those of the i-th oscillator in the i-th zone. Evidently, the perimeter of this contact zone grows with time. Ultimately, autowaves from the highest-frequency oscillator occupy the full area of the active zone, and the rest are suppressed.

If we carry out a reverse Fourier transform, successively assigning the time of local zone clearing, we can receive a precursory process signal as it goes through successive clearing stages. This signal will gradually

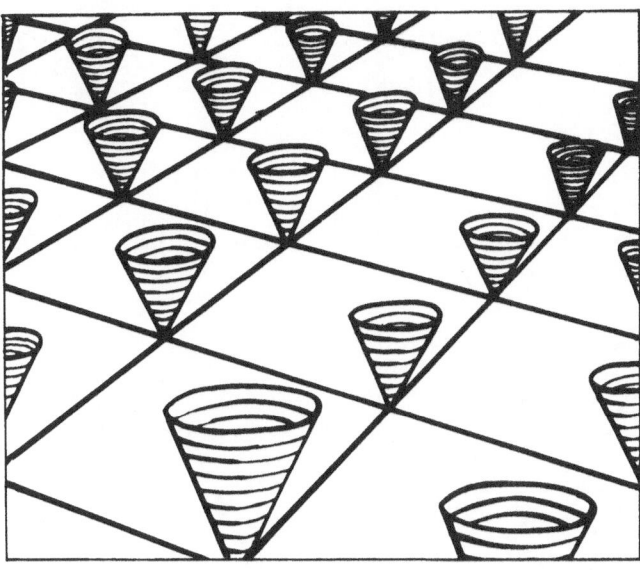

Fig. 3.14. A two-dimensional active autowave medium composed of auto-oscillators occupying all points of the coordinate grid. The divergent autowaves they emit over the medium surface can be interpreted as conic surfaces enlarging with time. The time axis is normal to the two-dimensional coordinate grid plane. What we have is a four-dimensional space with two coordinates in the plane and two on the surface of each cone. Thus the two-dimensional active medium has the same stereoscopic format as the four-dimensional world surrounding us (three spatial coordinates and one temporal) and can therefore be used to model it. Our brain, when looked upon as a pile of planar active autowave media, is five-dimensional. As far back as 1926, Fock and Klein suggested modifying the Schrödinger wave equation to use five instead of four variables. Solutions to this equation can be interpreted as waves propagating in the usual four spatiotemporal dimensions, but in the presence of a modulating gravitational or electromagnetic field. Thus the brain of living beings is complementary in dimension to the five-dimensional world of Fock–Klein and apparently tuned by evolution to interact with processes in a five-dimensional space

become free of noise and of decomposition harmonics (the latter actually modulate the process frequencies due to their closeness to high-intensity peaks).

The logic underlying the nonlinear dynamic filter described here is that of the active autowave medium. Such a filter is capable of self-clearing, and it differs from existing means of signal filtering in the Fourier region by its locally effective range. To create facilities for adequate frequency filtering one has to know beforehand what kinds of signals and noise one is dealing with. In our case, such knowledge is not necessary. The active autowave medium is self-alignable, therefore it

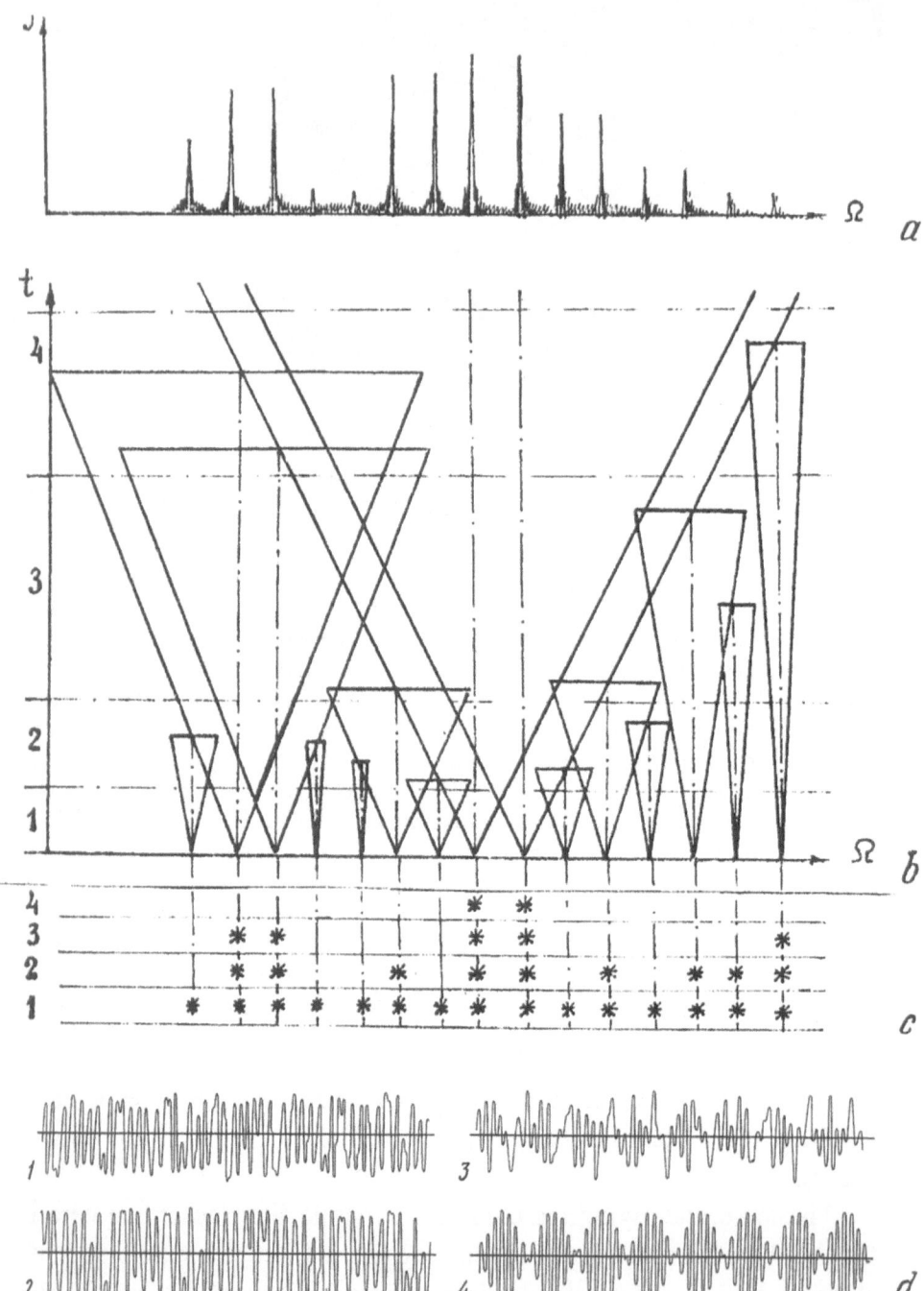

◄ Fig. 3.15a–d. A computer simulation example of the clearing of a one-dimensional process effected by the action of an autowave active medium.

(a) The Fourier image of the process: frequencies Ω are plotted along the horizontal axis and intensities J for the decomposition of the process into a Fourier series on the vertical axis. The Fourier image is projected onto a one-dimensional autowave active medium (diagram **b**). The medium has to synthesize the development of the auto-oscillator interaction. The auto-oscillator frequencies ω of space-time points in an autowave medium matrix are directly proportional to values of the intensity of the Fourier image harmonics.

(b) The process of area occupation by the highest-frequency oscillators (see Fig. 3.8). In one-dimensional space, conic surfaces (see Fig. 3.14) can be interpreted as triangular areas enlarging with time.

(c) Dots denote harmonics remaining after four successive clearing stages with the time intervals of diagram **b**. The clearing times are plotted on the vertical axis in diagram **b**.

(d) The four stages in the alteration of the real process with time. The conclusive regularity enabling a prediction to be made can be seen in panel 4: after approximately every 7 cycles a phase change occurs. The period (≈ 7 cycles) is the main modulating period of the model process. The process was shown in Fig. 3.12c ($2\omega_2/\omega_1 = 2 \cdot 3.45 \approx 7$). Clearing of a Fourier image by an autowave active medium allows the latent regularity of a process to be revealed, thereby enabling prediction of its development with time and in the space of the external world.

selects its own Fourier image clearing path using reference harmonics, and thus isolates the most important harmonic components characterizing the process. However, the prediction accuracy for this medium also has constraints influenced by the signal-to-noise ratio. Thus, for example, we can speculate that a prejudice is nothing but an erroneous prediction due to misinterpretation of noise as a signal.

Figure 3.15 shows a computer simulation of the clearing of a one-dimensional process by an active autowave medium. The only variable criterion for evaluating the quality of prediction, given the parameters of the active medium, is the time needed to clear the Fourier image. If the time chosen is too small, the active autowave medium is not finished with self-clearing. If the time allowed is too large, important signal components may be destroyed. One can see from the four diagrams in Fig. 3.15 how the clearing dynamics influence the pattern of the process. Figure 3.16 illustrates a scheme for a possible device establishing the sequence of actions involved in making a prediction by using an active autowave medium and a set of optical transducers [3.11].

Our neural structures probably operate using the same procedure, although the physical and chemical processes involved are different.

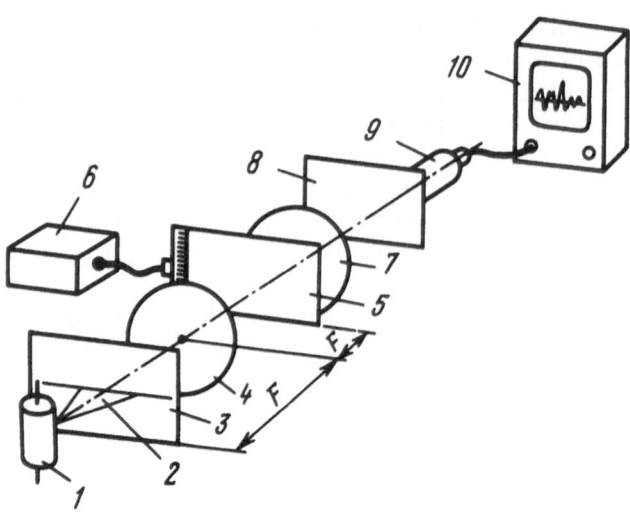

Fig. 3.16. Block diagram of a possible device for predicting the future behavior of a process from its past behavior by making a Fourier image of the function representing the process and clearing the image of noise and the harmonic components modulating its dominant frequency harmonics. 1 – laser source, 2 – laser beam, 3 – plane representing the function indicating the past behavioɪ of the process, 4 – first lens for transforming the past function into a Fourier image, F – focal length of the lens, 5 – Fourier plane where the Fourier image of the past function is formed by the first lens. An active autowave medium is located in this plane, which clears the Fourier image of noise and low-intensity harmonic components and performs a reverse tranform of oscillator frequencies into Fourier image intensities. 6 – power unit for the active autowave medium, 7 – second lens for effecting reverse transform of the Fourier image from the active autowave medium, 8 – plane representing the function indicating the future behavior of the process, 9 – device for recording the function, 10 – means to measure and visualize the possible future behavior of the process.

Matrices of brain auto-oscillators, however, are nothing like the homogeneous system discussed above. Initial differences in the values of the neurons' structural parameters have already been set genetically in the brain autowave medium. This is exactly how genetic memory of the past is embodied. The morphological diversity of these parameters together with variety in the modules they make up determine an a priori model representing average features of the environment in which the human brain has been evolving for the last 40 000 years. The initially homogeneous autowave medium with its auto-oscillators discussed here represents a brain structure free of any genetic past experience. The main conclusion we can infer from the study of the electrical activity within the cerebral cortex is that the brain is a particular kind of active autowave medium that has developed to predict processes evolving spatially and temporally both inside and outside us. This active medium is most favorably adapted to forecast events in the time domain of human existence. Thus, our failures to predict successfully, no matter whether they are due to haste or to forgetfulness, are errors for which we ourselves are to blame.

References

3.1 G.R. Ivanitskii: Rhythms of developing complex systems. Mathematics and Cybernetics 9. Moscow: Znanie, 1988 (in Russian)

3.2 G.R. Ivanitskii; V.I. Krinsky, E.E. Selkov: The cell's mathematical biophysics. Moscow: Nauka, 1978 (in Russian)

3.3 V.I. Krinsky, A.S. Mikhailov: The autowaves. Physics 10. Moscow: Znanie, 1989 (in Russian)

3.4 G.R. Ivanitskii: Autowaves within and outside us. In: Yearbook: Science and mankind, pp. 211–226. Moscow: Znanie, 1989 (in Russian)

3.5 G.R. Ivanitskii: Biological microdevices. Vestnik AN SSSR 3, 118–128 (1984) (in Russian)

3.6 N.N. Vsevolodov, G.R. Ivanitskii: Biological light-sensitive complexes as technical carriers of information. Biofizika 30(5), 883–887 (1985) (in Russian)

3.7 G.R. Ivanitskii, N. Vsevolodov (eds.): Light-sensitive biocomplexes and optical sensing of information. Pushchino: Inst. Biophysics Acad. Sci. USSR, 1985 (in Russian)

3.8 G.R. Ivanitskii (ed.): Biotechnology as a new computerization direction. Moscow: Nauka, 1990 (in Russian)

3.9 V.I. Arnold: On the instability of dynamical systems with many degrees of freedom. Doklady Acad. Nauk USSR 159(1), 9–11 (1964) (in Russian)

3.10 G.R. Ivanitskii, E.A. Goida, A. A. Deev et al.: Is the synchronization of cell division in early embryogenesis determined by external or internal factors? Biophysics 36(2), 358–361 (1991)

3.11 G.R. Ivanitskii, A.S. Kuniskii: Studying an object's microstructure by methods of coherent optics. Moscow: Energia, 1981 (in Russian)

4. Synergetics, Predictability and Deterministic Chaos

G. G. Malinetskii

4.1 Dynamical Chaos

Although we do not speak about it, we all know that the objectives of our science are, from a general human point of view, much more modest than the objectives of, say, the Greek science were; that our science is more successful in giving us power than in giving us knowledge of truly human interest.

Eugene P. Wigner

Since ancient times, the ability to predict was believed to be a divine right of sages and one of the main goals of the development of science. Foretelling the eclipses of the sun and the moon by priests was an object of awe and wonder. In writing biographies of famous Greeks and Romans, Plutarch presumably meant that by turning back to the past it is possible to look into the future of individual persons, states and races. However, he left it to the reader's own perspicacity to construct prediction algorithms.

The situation changed dramatically when laws governing nature were established mathematically. It appeared that the motions of celestial bodies could be computed using differential equations. The investigation and solution of these equations requires a good deal of effort and inventiveness and entirely new mathemaical instruments, but is not in principle an unfeasible problem. The idea excited scientists and engrossed their minds for a long time. Indeed, an extensive domain of investigation opened out in which scientific prediction seemed possible. The only apparent hindrances to successful forecasting were of mathematical origin and seemed likely to be eliminated with time.

The brilliant French mathematician Pierre Simon de Laplace, whose works did much to overcome these difficulties, believed that the main goal of contemporary and future science was to derive corollaries to the laws of Newtonian dynamics. When asked by Napoleon why there was no place left for God in his conception of the universe, the mathematician replied with dignity, "I have no need of that hypothesis". What an admirable belief in the unbounded capabilities of the mind and in the rationality inherent in the world!

Since Laplace's time, the advance of science is commonly presented as having been a succession of triumphs. Momentous discoveries were made that unfolded new possibilities before man, allowed more precise description of various phenomena, and hence improved the quality of prediction [4.1].

However, there is a contrary, and equally justifiable, view of the development of science. In fact, the majority of fundamental theories that have altered previous directions of research (the paradigm-altering theories, as the historian of science Thomas Kuhn called them) have also brought with them the realization of new limitations. These limitations are mainly associated with discovering that certain objectives cannot be pursued by a scientist even in principle.

For example, given a system that is perfectly described by Newtonian laws, we cannot alter the equations simply by reversing the sign of the time variable. A motion picture film recording the development of such a system can be run both forward and backward. Viewing the motion picture, it is impossible to determine from the laws of mechanics which of the two versions is realizable in nature. However, with the advent of thermodynamics and statistical physics, the notion of irreversibility found its way into the natural sciences. It became clear in general why a process cannot be run backward in nature, as well as why perpetual motion is impossible.

Quantum mechanics has provided a dramatic proof that it is fundamentally impossible to make simultaneous measurements of both the position and momentum of a particle with arbitrary precision. There are many other pairs of (complementary) quantities that cannot be simultaneously measured with precision. The theory of relativity has similarly revealed numerous insurmountable barriers. In effect, many of the most important developments in science have narrowed the scope of the questions that we can put to nature.

Among the particularly important achievements are recent discoveries related to predictability and so-called dynamical chaos. These made researchers aware of yet another barrier. It was found that a fair proportion of relatively simple mechanical, physical, chemical, and ecological systems are inherently unpredictable over long time periods. Presumably, predictability for short terms and unpredictability for long terms is characteristic of many phenomena studied in economics, psychology, and sociology.

The forefront of fundamental investigations is commonly associated with tremendous astrophysical scales or with the microstructure of matter. Studies of dynamical chaos proved that certain paradoxical features that have only recently become subject to inquiry are found in phenomena that are fairly well described by classical mechanical laws.

A new science, nonlinear dynamics, has filled in many gaps in the understanding of predictability. Before summarizing the essentials of this field, we mention some differences between the new, developing concepts and the old ones.

It was previously thought that there are two classes of objects in nature. The first are deterministic objects whose behavior is predictable for any desired term. The others are stochastic (that is, random) ones, which are the subject matter of probability theory. Typical cases of stochasticity are coin flipping and dice casting, where the results have nothing to do with the history of prior behavior. There is no determinism here that enables forecasting, and instead we have to treat such processes in terms of statistical characteristics, such as mean values, dispersions, probability distributions, and so on.

In the last twenty years, a third, previously unknown class of objects has been proved to exist. This class comprises objects which are formally deterministic, since from a knowledge of their current state we can predict their state for more or less long time periods. Actually, however, prediction is possible only for a limited period of time. The fact is that due to a very sensitive dependence on initial conditions, any change in the accuracy with which these initial conditions are given, however small it might be, produces a growing degree of uncertainty, so that beyond a certain time limit prediction becomes impossible. The systems's behavior beyond that limit is chaotic. Again, we can only treat such processes in statistical terms. The systems in question are found in hydrodynamics, laser physics, chemical kinetics, astrophysics and plasma physics, geophysics, and ecology. The fields in which predictability is limited are vast indeed.

However, in many cases the barrier is not merely a reason to become disillusioned. It also allows one to see either promising prospects or the true scope of the problems at hand.

This chapter is addressed to the topic of dynamical chaos. The novel concepts of nonlinear dynamics presented here will probably be of interest to readers engaged in many different fields of activity. With a view to making the main ideas understandable to a nonspecialist audience, I have tried to avoid using too many formulas. In fact, the role of formulas here appears to be much like that of gestures in conversation. Keeping track of them makes the verbal content more descriptive.

4.2 Nonlinearity and Open Systems Behavior

> Thus reason draws things
> out of chaos,
> And orders them according
> to causal connections, time
> and space,
> Consolidating them
> by the power of numbers.
>
> *M. Voloshin*

The use of ordinary differential equations as mathematical models (to use the language of modern science) was considered by Newton to be one of his most important achievements. The equation

$$\frac{d\mathbf{x}}{dt} = \mathbf{f}(\mathbf{x}, \lambda), \quad \mathbf{x}(0) = \mathbf{x}_0, \quad \mathbf{x} = \{x_1, \ldots, x_N\} \tag{4.1}$$

governs the rate of change of vector $\mathbf{x}$ with respect to time, where λ is an external parameter and $\mathbf{x}$ consists of N components and has initial value $\mathbf{x}_0$. This vector can define the velocity and position of material points, the set of velocities and positions of planets in the solar system, a set of currents and voltages, and a good many other quantities.

It now remains only to wonder at the elegance of Newton's laws, which appear to be so simple. If we perform a few relatively easy calculations starting from equation (4.1) we can predict the motions of celestial bodies and the existence of unknown planets, or the structure of a resonator's electromagnetic field, or the spectrum of radiation from a hydrogen atom.

The classical mechanics paradigm based on the use of equation (4.1) admitted two central ideas. The first can be illustrated by the following example. Setting $\mathbf{f} = \{x_2, -\omega^2 x_1\}$, we can rewrite (4.1) as

$$\dot{x}_1 = x_2$$
$$\dot{x}_2 = -\omega^2 x_1 \, .$$

Elimination of x_2 reduces these equations to the form

$$\frac{d^2 x_1}{dt^2} + \omega^2 x_1 = 0 \, .$$

An amazingly large number of different systems satisfy this equation, ranging from the oscillation of a weight on a string to the changing populations of predators and their prey in some area. The solutions to the equation take the form

$$x_1 = A \sin \omega t + B \cos \omega t \, .$$

An important point is that the function $\mathbf{f}$ is linearly dependent on unknowns x_1 and x_2 (that is, $\mathbf{f}(\alpha \mathbf{x} + \beta \mathbf{y}) = \alpha \mathbf{f}(\mathbf{x}) + \beta \mathbf{f}(\mathbf{y})$). Equations of this sort have very simple solutions involving sines, cosines, and exponential functions. The gifted American physicist Richard Feynman once noted that the vast majority of theoreticians spend most of their working lives solving linear equations and that the main reason for this is that linear equations are in fact the only class of mathematical models that can be analyzed in detail. Linear equations are like the famous streetlamp under which scientists are especially fond of searching for their keys [4.2].

The natural course of reasoning seems to be this. By solving one equation describing pendulum motion we obtain some very simple and useful functions that can be employed for solving many other problems. We now look for other linear equations, since their solutions probably lead us to other useful functions. It was this course that mathematicians followed. Truly remarkable functions were introduced by Hermite, Laguere, Legendre, Bessel, and Jacobi.

The other idea is connected with the use of asymptotic methods. Its essence is very simple, too. Suppose we can solve equation (4.1) for function $\mathbf{f} = \mathbf{f}_0(\mathbf{x})$, where $\mathbf{x} = \mathbf{y}_0(t)$, but we are interested in finding a solution to $\mathbf{f} = \mathbf{f}_0 + \varepsilon \mathbf{f}_1$, where ε is some small quantity (in the language of mathematics, a small parameter). Then it is only natural to look for a solution in the form of a series

$$\mathbf{x}(t) = \sum_{k=0}^{\infty} \varepsilon^k \mathbf{y}_k(t) \, .$$

For each function $y_1, y_2, \ldots$ we shall obtain a special equation, and they will all be much simpler than the initial one. Furthermore, it is possible that even the first few members of the series will enable us to determine the solution with high precision. This approach has proved to be very efficient in many situations and is still being developed intensively today. Cases where it can be used to good effect range from celestial mechanics, the most interesting problems of which have been solved with its aid, to quantum electrodynamics and the general theory of relativity.

The two above ideas, standard solutions for linear equations and the use of asymptotic methods, underlie the conception of determinism that has given rise to our belief in the unbounded capability of the human mind to reveal all the fundamental laws of nature and to bring to light the entire set of their corollaries.

However, at the turn of the century it became clear that the scope of linear equations whose solution involves the introduction and detailed investigation of certain special functions (sine analogues) is not at all large. A comparatively small set of problems lent itself to study by asymptotic analysis.

Using numerical methods to search for a solution without knowing exactly what it should look like, even given the availability of powerful computers, is practically the same as searching for lost keys in the dark. A new conception was needed.

It was provided at the beginning of the century by the leading French mathematician Henri Poincaré. He radically altered the statement of the problem. According to Poincaré, in many cases one should focus not on all possible solutions, but rather on a certain limiting (asymptotic) state which the system reaches after very large times (formally, when variable t tends to infinity).

This statement of the problem brought forth a new paradigm, now the subject of intense development. The point about this idea is that not only has it created a fundamentally new type of mathematics, which has ceased to be just the science of numbers and geometrical figures, by giving rise to such new fields of research as topology, catastrophe theory, bifurcation theory, and so on, where qualitative rather than quantitative properties of objects are the object of study, but also, and even more importantly, it has made it evident that a host of equations of diverse form governing a variety of processes are inherently related.

An example will give a better insight into the situation. Let us assume that equation (4.1) governs some open system (that is, one capable of exchanging energy, matter, and so on with the environment). We can also say an open system is one with energy dissipation, which constitutes a class to which the majority of systems related to ecology, chemistry, or radio engineering belong. The class contrasts with that of traditional energy-conserving models, as in celestial mechanics. A remarkable point about open systems is that whatever function $\mathbf{f}$ is taken, when $N = 2$, we eventually observe just two different behavior patterns, illustrated by Figs. 4.1 and 4.2.

Let equation (4.1) describe the trajectory of a point governing the system's state in space $\{x_1, \ldots, x_N\}$, known as phase space. The trajectory itself is called a *phase trajectory*. The behavior of the system when $t \to \infty$ is determined by an attracting set, or an *attractor*. Take any component of vector $\mathbf{x}$, say, $x_1(t)$. After some transient process (from $t = 0$ to $t = t_1$ in Fig. 4.1) it reaches a steady state, as do other vector components. The attractor shown in Fig. 4.1 is a point in the phase space. In the language of mathematics, this is a stable singular point of the system (4.1). The second possible outcome is illustrated by Fig. 4.2. Following some transient condition, a periodic oscillation occurs. The attractor in this case is a closed curve called a *periodic orbit*, or a *limit cycle*.

That's all. No other qualitative behavior patterns occur for an immense variety of functions $\mathbf{f}$ providing bounded solutions are to be found. Isn't that surprising?

Suppose that we want to predict the behavior of a system containing a stable singular point or a limit cycle. Assume that there is a small error $\mathbf{d}(0)$ in our knowledge of the system's initial condition (see Fig 4.3), so that $\mathbf{x}'(0) = \mathbf{x}(0) + \mathbf{d}(0)$. Assume we also know the governing equation (4.1). We now track both the trajectory $\mathbf{x}(t)$ (referring to the real process) and the trajectory $\mathbf{x}'(t)$ (referring to the prediction). The divergence from the real state $\mathbf{d}(t) = \mathbf{x}'(t) - \mathbf{x}(t)$ appears not to be growing; it may even diminish. In this way, a whole class of initial conditions will be able to reach the steady state. We call this phenomenon *forgetting initial data details*.

At this point, we shall introduce some numerical characteristics of the rate at which nearby trajectories tend to approach one another. Take two

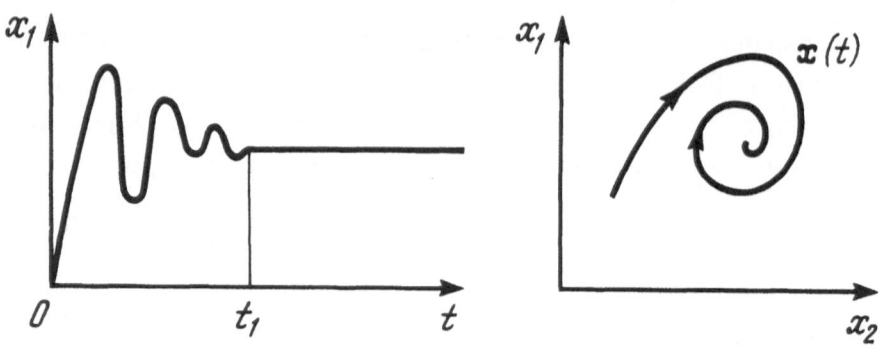

Fig. 4.1. A typical behavior pattern for the case when the attractor is a stable singular point. The plot of $x_1(t)$ and a trajectory in phase space

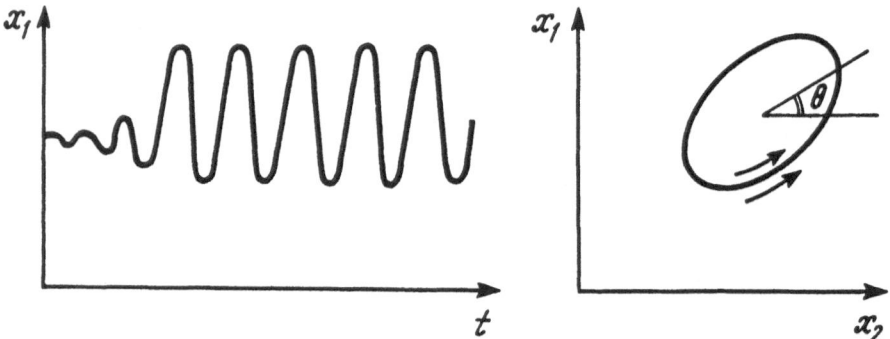

Fig. 4.2. A typical behavior pattern for the case when the attractor is a limit cycle

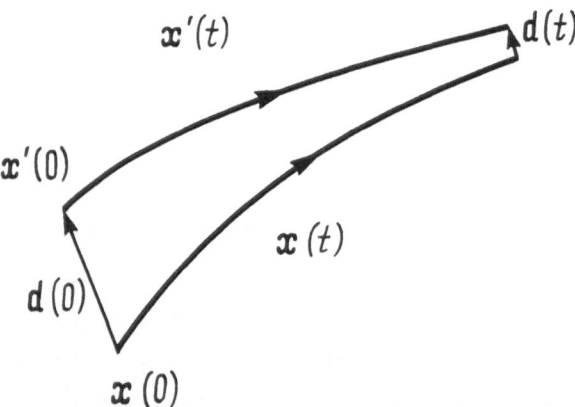

Fig. 4.3. For a predictable system containing a stable singular point or a limit cycle, $d(t)$ diminishes and nearby trajectories $x(t)$ and $x'(t)$ converge

nearby trajectories $\mathbf{x}(t)$ and $\mathbf{x}'(t)$ and see how the length $d(t) = |\mathbf{x}'(t) - \mathbf{x}(t)|$ (i.e., simply the length of the vector $\mathbf{d}(t)$) will change. If the trajectories converge, then $d(t) \approx e^{\lambda t}$, $\lambda < 0$. The quantity λ is called the Lyapunov exponent and is formally defined as

$$\lambda(\mathbf{x}(0), \mathbf{d}(0)) = \lim_{t \to \infty} \lim_{d(0) \to 0} \left[\frac{1}{t} \ln \frac{d(t)}{d(0)} \right].$$

If it is positive, the Lyapunov exponent actually gives the mean rate of divergence of two infinitely close trajectories; if negative, it gives the rate of convergence.

It would appear that choosing different points $\mathbf{x}'(0)$ in the neighborhood of $\mathbf{x}(0)$ as well as different $\mathbf{x}(0)$ should give us different values. However, in 1968 the Soviet mathematician Oseledets proved that under some additional conditions all the points $\mathbf{x}'(0)$ and $\mathbf{x}(0)$ in the neighborhood of a strange attractor of an N-dimensional system (4.1) yield one and the same collection of N Lyapunov exponents $\lambda_1, ..., \lambda_N$ [4.3]. Their meaning is rather simple. The variation of length $|\mathbf{x}'(t) - \mathbf{x}(t)|$, where $\mathbf{x}(t)$ and $\mathbf{x}'(t)$ are infinitely close trajectories, varies according to the quantity $\exp(\lambda_1 t)$ given by the λ_1 exponent. The area of the parallelogram with vertices $\mathbf{x}(t)$, $\mathbf{x}'(t)$, $\mathbf{x}''(t)$ (Fig. 4.4) varies according to quantity $\exp[(\lambda_1 + \lambda_2)t]$. The volume swept out by the parallelogram varies according to quantity $\exp[(\lambda_1 + \lambda_2 + \lambda_3)t]$, and so on. Since we consider N-dimensional systems that have attractors, which means their N-dimensional volume decreases, we have $\lambda_1 + ... + \lambda_N < 0$. For a stable singular point, $\lambda_i < 0$, for all $i = 1, ..., N$. For a limit cycle, $\lambda_1 = 0$ and $\lambda_i < 0$, for $i = 2, ..., N$ [4.4].

The problem is even simpler if we are interested only in the steady-state conditions. Indeed, let us wait for a sufficiently long time for the steady state to be reached. If the process develops as shown in Fig. 4.1, we should only need to focus on limiting values, and it should suffice to work on the algebraic system $\mathbf{f}(\mathbf{x}) = 0$ instead of solving equation (4.1). In the case depicted in Fig. 4.2, we should introduce a coordinate indicating the position of the point in the cycle, for example, angle θ, and solve a simpler equation governing the variation of θ instead of equation (4.1). This angle governs all the other variables: $x_1 = x_1(\theta), ..., x_N = x_N(\theta)$. Thus, with such an idealized formulation of the problem, we have every possibility to produce a 'global forecast', that is, a forecast for any desired term and with very high accuracy.

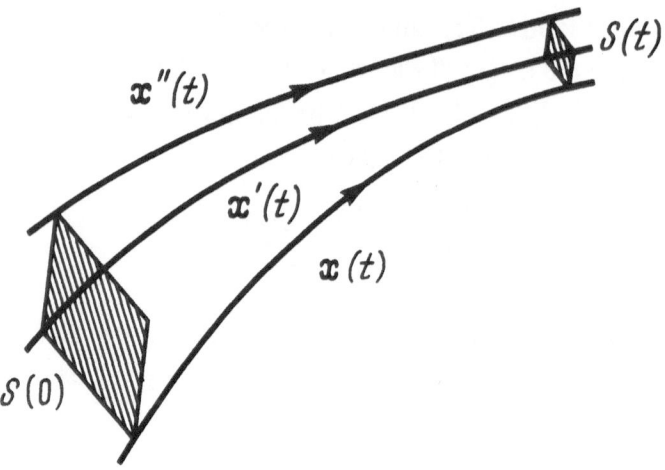

Fig. 4.4. The sum of the first two Lyapunov exponents characterizes the variation of the area $S(t)$ of the parallelogram with time

For $N > 2$, an attractor which presents a 2-frequency state is possible. This attractor is called the invariant torus and has the form of a donut with a trajectory coiled around it (Fig. 4.5). In the simplest case involving an invariant torus, variations of function $x_1(t)$ resemble beats, where $\lambda_1 = 0$, $\lambda_2 = 0$, and $\lambda_i < 0$ for $i = 3,...$ Consequently, forecasting is also possible.

Naturally, all this is simple only in principle. When analyzing real systems, the main effort involved is setting up function $\mathbf{f}$, which in effect is equivalent to constructing the mathematical model of the phenomenon. It is far from easy in most cases to measure the initial state $\mathbf{x}(0)$ and the current state $\mathbf{x}(t)$ of the system.

Moreover, the equation at hand can have a number of attractors, as illustrated by Fig. 4.6. The attractors in this figure are stable equilibrium points O_1 and O_2. Each has its own basin of attraction, A and B, respectively. All trajectories starting in A tend to O_1, while those starting in B tend to O_2.

Assume that we know all possible attractors that can conditionally be called the end states of the evolution process in a system, as well as the initial state $\mathbf{x}(0)$. This is still insufficient if we want to know how the system will eventually evolve or if we want to control the process. What we need to know in addition are the attraction basins of the different attractors.

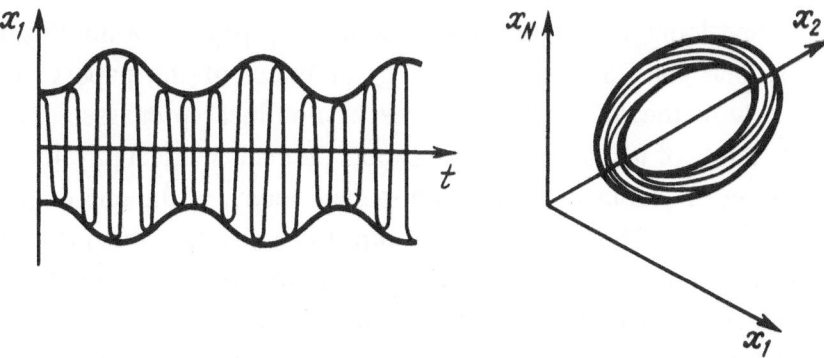

Fig. 4.5. Possible appearance of an invariant torus in phase space and the corresponding plot of $x_1(t)$

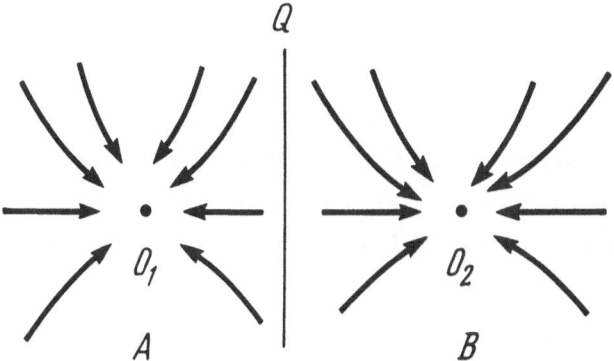

Fig. 4.6. Schematic drawing to show that a slight variation of initial data at the boundary Q (here a symmetry axis) of the attraction basin may qualitatively alter a system's behavior

We can find a very nice attractor O_1 and then wait eagerly for the system to reach it. But if the system's initial state happens to be far away from basin A, then our attractor O_1 is unattainable. Such problems are often encountered in ecology and chemical technology. There might be other circumstances, however, under which point $\mathbf{x}(0)$ is at an attraction basin boundary. Then we can choose the end state of the evolution of the process by slightly varying $\mathbf{x}(0)$.

Attractors, limit cycles, and attraction basins are key terms in the language of nonlinear dynamics, and scientists in various disciplines use them with increasing frequency to formulate their findings. The role of both the language itself and its associated images has grown significantly. If one looks through a journal containing papers on nonlinear dynamics, one can easily see that a great many of the papers are concerned with the qualitative features of nonlinear processes. The contrast in terms of language between papers on mathematical physics at the beginning of this century and such papers today, at the end of the century, is striking. The cause of this contrast is apparently that most of the problems modern scientists have to tackle are nonlinear. We don't know much about the properties of the nonlinear world.

With the advent of computers, a new technique has been developed to research this world – the *computational experiment* or *simulation*. This has opened up new possibilities for testing bold and unusual hypotheses and for computing possible alternatives in advance and in great detail before turning to a field experiment. The result has been a growth in the importance of intuition, analogies, and images offered by mathematical physics, or sometimes by quite remote disciplines. We may perhaps compare the present state of development of scientific research with that of science in antiquity. Like the ancient Greeks who made guesses about how the world was constituted based on myths, analogies, and peculiar images, and thus sometimes achieved marvelous results, modern scientists are now seeking to understand the complexity of the world in terms of the simplest nonlinear models.

4.3 Synergetics and Order Parameters

Note first of all that any generalization implies believing to a certain extent in Nature's uniformity and simplicity. The assumption of uniformity presents no difficulty ... The question therefore is not whether Nature is uniform, but rather in what manner it is uniform.

Henri Poincaré

When we read books about science and knowledge we often come across the image of a giant library built as a labyrinth. The person in the library is unable to become acquainted with even a small portion of the library's contents or to distinguish between the truths and the delusions stored in the labyrinth. Only by lucky chance can the mysteries of the library be revealed. The feelings of the owner of such a treasury who has no idea of how to dispose of his property is subtly rendered by such masters of fiction as Jorge-Luis Borges, Herman Hesse, and Umberto Eco.

The causes of the increasing anxiety many researchers feel for the fate of science are precisely formulated by Eugene Wigner, one of the originators of quantum mechanics, in his essay 'Limits of Science'. He starts by defining what can be regarded as scientific activity: "I would say that a store of knowledge can reasonably be called 'our science' if there are people who are competent to learn and use any part of it, who would like to know each part of it even if they realize they cannot, and if one has good assurance that the parts are not contradictory but form a whole". He goes on to make a pessimistic forecast: "It is not difficult to imagine a stage in which the new student will no longer be interested, perhaps will not be able any more, to dig through the already accumulated layers in order to do research at the frontier ..." [4.5]

There are, of course, two stabilizing strategies that can be used when tackling these problems, namely, to eliminate whole fields of science from the active area of research, and to find shorter paths through the labyrinth of learning: "the more we dip into the idea of the discovery the better we can explain it".

Naturally, the importance of these strategies has not diminished in the time that has elapsed since Wigner wrote his paper. However, many workers presently pin their hopes on the development of interdisciplinary approaches that might enable us to gain a better insight into the organization of the world and to reveal the commonality of many natural phenomena through the body of hypotheses, methods, and techniques that apply across the different disciplines. Such approaches may eventually play the role of Ariadne's thread in the labyrinths of science.

Synergetics is the science of the general laws of order that arise in open nonlinear dissipative systems. It has considerably altered our view of many phenomena. Within synergetics, many complex systems characterized by a large (sometimes infinite) number of degrees of freedom are found to behave quite simply. Conversely, the behavior of a simple system with as few as three degrees of freedom sometimes appears to be very complex, indeed chaotic. Both of these features are closely connected with the problem of predictability [4.6–8]. To fix our ideas, let us concentrate upon the key concept of synergetics, that of *self-organization*. As a reminder, the *number of degrees of freedom* of a system is determined by the number of quantities that need to be given in order to characterize the state of the system. For instance, for objects governed by ordinary differential equations (4.1), this number is just N.

Nowadays, however, scientists are increasingly interested in studying problems involving variations in time and space, when formally $N = \infty$.

Let us examine the example of variations in the distribution of reactants in a chemical reactor. In this case, instead of the N-dimensional vector **x** in equation (4.1) we have to deal with a collection M of functions $\mathbf{u}_i(r,t)$, $i = 1,\dots,M$, describing the concentration of the i-th reactant at the point with coordinate r at time t. For simplicity, we denote all these functions as vector $\mathbf{u}(r,t)$, where of course the vector has different values at different points r and times t.

Assuming that the behavior of the reactants is determined by chemical reactions and diffusion processes, the equation governing the evolution of $\mathbf{u}(r,t)$ can be written symbolically as

$$\mathbf{u}_t = \hat{D}\mathbf{u}_{rr} + \mathbf{F}(\mathbf{u},\lambda), \quad 0 \le r \le l. \tag{4.2}$$

Here $\hat{D}\mathbf{u}_{rr}$ describes diffusion processes (since matrix $\hat{D}$ is composed of diffusion coefficients) and $\mathbf{F}(\mathbf{u},\lambda)$ gives the kinematics of chemical reactions. The length of the reactor is l, and λ is an external parameter. Solutions of the equation are functions of variables r and t. Since partial derivatives (with respect to r and t) appear in (4.2), it is a partial differential equation.

Physicists and chemists concerned with process simulation, as well as engineers concerned with applying the theory of elasticity and researchers in hydrodynamics, are all well acquainted with partial differential equations like (4.2). Since the time they were first formulated by Laplace, they have held a firm position in the bedrock of natural science.

When function **F** is linear, equation (4.2) is relatively easy to solve. But if **F** is nonlinear, the equation – generally known as a *reaction–diffusion system* – falls into an area where intense research is under way using the latest methods. There have been thousands of studies devoted to such systems.

The introduction of $\mathbf{u}(r,t)$ is in itself a bold and substantial step. Asked to give an intuitive image of an electromagnetic field, Richard Feynman said 'I have no picture of this electromagnetic field that is in any sense accurate' [4.2]. To represent the field, one assumes that six defined numbers characterize the electric and magnetic fields at each point in space. The situation is much the same in our case, where $\mathbf{u}(r,t)$

is actually a set of M time-varying numbers at each point in space. One would like an intuitively simpler way to represent the field.

Synergetics offers some possible ways, one of which is as follows. Let us return to the subject of the attractor and equation (4.1). At $t \to \infty$ we had a rather simple description. For example, when a system reached the limit cycle, its state was defined by one quantity, the point's position in the cycle (angle θ in Fig. 4.2). All the other variables were merely functions $x = x(\theta)$ of that quantity.

It is the same with problems concerning partial derivatives. Not all degrees of freedom play similar roles. With a nonlinear dissipative system, it is usually possible to separate a limited (and sometimes small) number of variables, to whose values other variables 'adjoin'. Such central variables are sometimes called order parameters. By way of example, let us consider function $u(x)$ defined over the interval from 0 to l. We can expand it into harmonics (or, as mathematicians say, develop it as a Fourier series) and determine each harmonic's amplitude.

Alternatively, we can present function $u(x)$ as

$$u(x) = \sum_{m}^{\infty} C_m \cos\left(\frac{\pi m x}{l}\right).$$

The idea of this operation will be readily apparent if we recall the school experiment with a triangular glass prism. A ray of white light passes through it to give a beautiful color spectrum. Actually, the prism decomposes light into various wavelength components. In much the same way, this effect is repeated many times in TV and radio sets.

Another elementary analogy is a two-component vector expansion into x and y components. Within a function space, these components are functions $\cos(\pi m x / l)$, $m = 0,1,2,...$, which are infinite in number. The projections of each of them are determined by the amplitudes C_m of the harmonics.

Let function $u(x)$ have a complicated shape (Fig. 4.7a). Its behavior shows no sign of regularity or order. Many harmonic amplitudes are comparable. In contrast, the smooth function $u(x)$ in Fig. 4.7b behaves simply, so one can see its regularity, that it is close to a periodic function. Defining just some of the function's harmonic amplitudes suffices to render its profile. If the number of large-amplitude harmonics falls off as the process develops, this leads to a certain ordering in the system and enables its self-organization. What we eventually get are *dissipative*

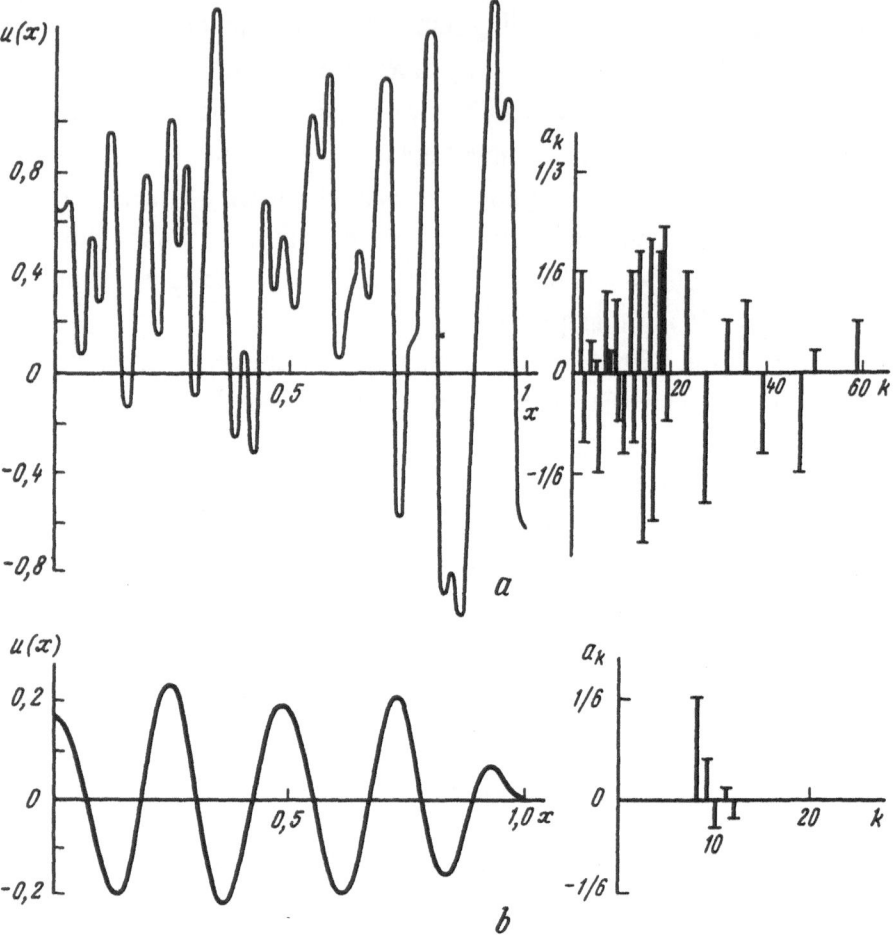

Fig. 4.7. Function $u(x)$ and its Fourier coefficients (the more jagged the function the greater the number of coefficients with close amplitudes)

structures (the name emphasizes the importance of dissipative processes in their formation).

The term 'dissipative structure' was introduced by the Belgian researcher Ilya Prigogine. Work by his scientific school has given a clearer insight into the connections between such structures and some concepts from thermodynamics. The studies have also stimulated both theoretical and experimental investigation of open system ordering. This is what Nicolis and Prigogine say about structures arising from disorder: "... both divergence from equilibrium state and nonlinearity may cause ordering of a system. A highly nontrivial relation may exist between order, stability, and dissipation. To better define this relation, we call the orderly configurations external to the thermodynamic branch stability domain dissipative structures... These can exist remote from the equilibrium state only due to a considerable flow of both energy and matter. The dissipative structures exemplify a striking ability of the non-equilibrium state to be a source of order" [4.7]. By the 'thermodynamic branch stability domain' is meant a domain of parameters that lends itself to study by the methods of nonequilibrium thermodynamics.

We can simplify our description of complicated systems by using the new paradigms offered by self-organization and dissipative structures. That is to say, we can now interpret complicated phenomena using plainer language.

Some similarities to images inherent in the most simply organized systems can be found in distributed systems (i.e., systems depending on spatial coordinates). In a system of form (4.1), the simplest attractor is a stable singular point: as $t \to \infty$, values $\mathbf{x}(t)$ tend to a constant. Its analogue is the time-independent solution to equation (4.2), for example, a stationary dissipative structure. A graphic example of such a structure is shown in Fig. 4.8. What we see is a typical steady-state distribution of reactants in some mix containing two substances, X and Y, $\mathbf{u} = (X,Y)$. At each point in the mix, components of vector $\mathbf{u}$ tend to constant values, $\mathbf{u}(r,t) \to \mathbf{u}'(r)$. As this takes place, $\mathbf{u}'(r)$ itself may have a rather complicated shape. Still, it is nothing but a stable specific point within some phase space. We can just as easily find the limit cycle analogues. One of them, developing in a certain model system important for studying equation (4.2), is illustrated in Fig. 4.9. Isn't it beautiful?

However, what is really impressive is not the comparison of different types of order and system attractors, but their evolution as they follow changing problem parameters.

One of the most remarkable manifestations of the nonlinear world is that of solution branching, or *bifurcation*. The latter term denotes variation of the number or stability of solutions of a certain type. A simple bifurcation is presented by an ordinary quadratic equation:

$$ax^2 + bx = \lambda \quad (a > 0).$$

We now alter λ. At $\lambda < \lambda_0 = -b^2/4a$, it has no solutions at all. At $\lambda = \lambda_0$ the equation has one, and at $\lambda > \lambda_0$ two solutions. Bifurcation takes place at point $\lambda = \lambda_0$, as illustrated in Fig. 4.10, where the abscissa represents λ values and the ordinate plots roots of the equation. Such figures are called bifurcation diagrams.

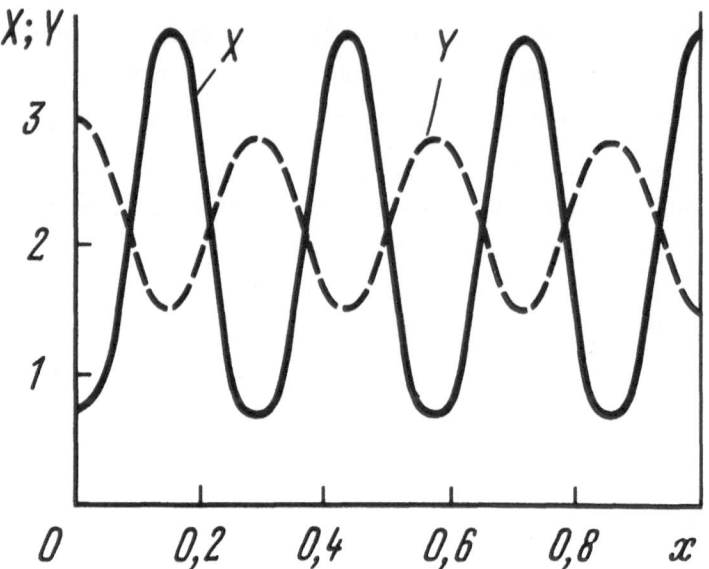

Fig. 4.8. A typical behavior pattern for a stationary dissipative structure. Functions $X(r,t)$, $Y(r,t)$ define concentration distribution within a model describing a chemical reaction

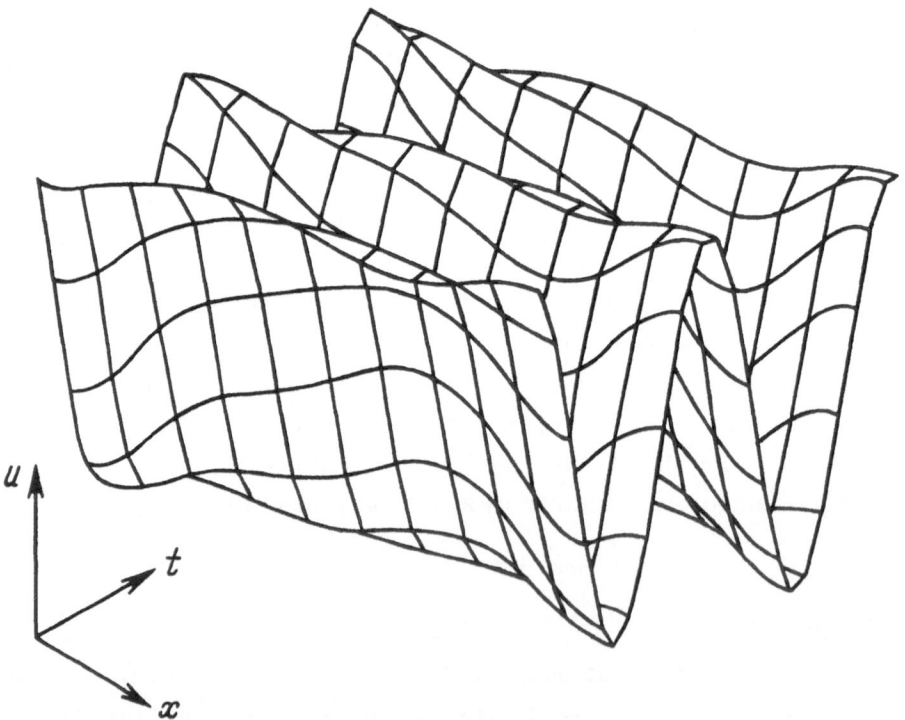

Fig. 4.9. Solution to the case where a periodic mode occurs in a model oscillatory chemical reaction system

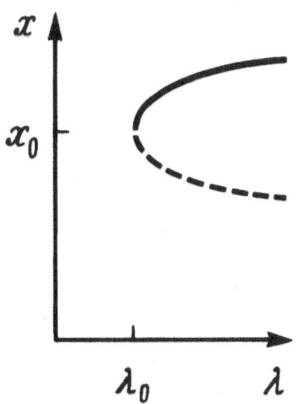

Fig. 4.10. The plot of $x(\lambda)$. Stable and unstable solutions are on the solid-line and the dashed-line branches, respectively

In the case of a chemical reactor in which the concentration of substance x is governed by the differential equation

$$\frac{dx}{dt} = ax^2 + bx - \lambda ,$$

the steady states (with a time-independent x) are also described by the same quadratic equations. Here we have to control both the number of solutions and their stability. As Fig. 4.10 shows, only one solution out of two is stable.

A striking feature of nonlinear dissipative systems, which incidentally has made possible the new science of synergetics, is that identical bifurcations occur in the original systems, in models connecting order parameters, and in the simplest equations. Some very simple examples are shown in Fig. 4.11. The fact that there is only a small number of elementary bifurcations demonstrates the inherent unity of nature. Realization of this fact has given rise to many areas of research [4.7–4.10], among them the theory of catastrophes [4.11]. The extensive list of research fields in which these results are applied (stability of vessels, mechanics of fluids, optics and scattering theory, thermodynamics and phase transitions, laser physics, ecology and embryology, social modeling, and many more) suggests inherent interrelations between a variety of natural phenomena.

Even if the bifurcations of a system are the only information available about it, this information can be very helpful. A simple sociological example will give a better insight here.

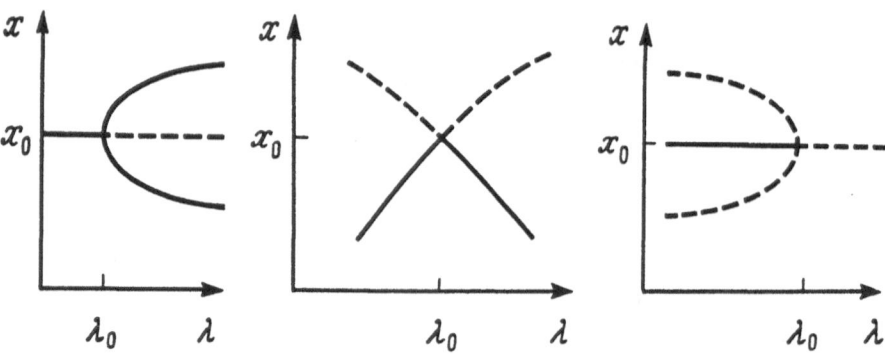

Fig. 4.11. Typical bifurcation diagrams for the case when at least one of the second-derivative functions $F(x,\lambda)$ at point (x_0,λ_0) differs from zero

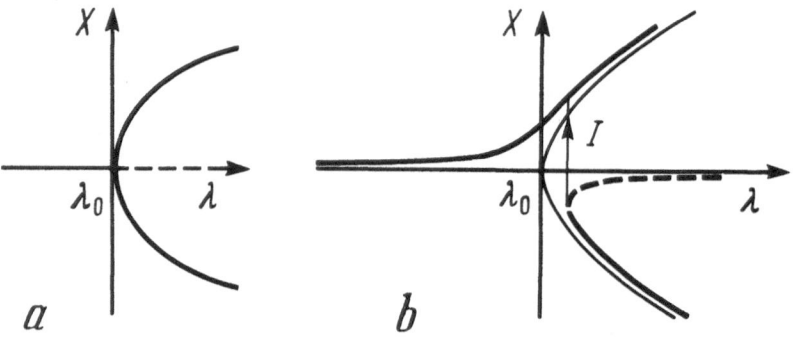

Fig. 4.12a,b. Typical behavior of equilibrium states in the model of public opinion formation: (a) the case where neither of the opinions is initially preferred, (b) the case where one of the opinions is initially preferred. Stable and unstable equilibrium states are indicated by solid and dashed lines, respectively

Suppose that a community is voting on some question. The system is characterized by the difference x between the number of voters in favor and the number against. Parameter λ indicates the degree to which the issue is 'hot', or significant. Figure 4.12a shows a bifurcation diagram typical of such circumstances. Let us see what is going on, assuming that neither of the two opinions is initially preferred. For $\lambda < \lambda_0$, the community makes its choice (by a majority vote). The system moves along either the upper or the lower branch. A typical pattern for the case where one of the opinions supersedes the other is shown in Fig. 4.12b. We may try to alter the choice of the community for a short period of time using limited means (say, one casting vote).

If we make no attempt to alter the situation, the choice of the community will vary as shown by the upper curve in Fig. 4.12b. For $\lambda \ll \lambda_0$, all our attempts will be in vain, since the issue is neglected. For $\lambda \gg \lambda_0$, they will also be in vain, since public opinion is settled. However, in the neighborhood of point λ_0 a small influence may determine the future fate of the community. Such a situation is typical of many systems studied in synergetics. All this invites meditation on the role of individual personalities in history.

Scientists in many fields of knowledge make use of extrapolation. In our present example, extrapolation gives us something like 'tomorrow will be much the same as today'. However, the lower curve in Fig. 4.12b shows that this is not always the case. Even if public opinion is very polarized, the parameter λ may fall in value. Assume that the system's state is defined by points on the lower branch of the curve. If a small change in λ occurs when it is near λ_0, public opinion changes abruptly, and a jump from the lower to the upper branch occurs. Public opinion is changed drastically, by an overwhelming majority. Examples of such jumps, analogous to 'phase transitions' in physics, are quite common in real life and are now the subject of intense study.

Let us draw a conclusion. Synergetics offers a method for studying complex systems as follows. First, determine the order parameters and their interrelations, and second, analyze all the equations that arise. If the equations still appear to be too complicated, look for new order parameters they may contain and construct new models. By constructing and analyzing a *model hierarchy* in this way, one can get down to the initial system. One can go further with this approach and try to forecast the system's behavior [4.12].

This strategy is a rather general one, for it is not restricted to any specific types of systems or equations and is therefore very attractive for workers concerned with a variety of problems. But this general character is also what undermines the efficiency of the strategy. It is not quite clear how to find order parameters for a given system and measure them. However, if they are available, one can easily describe very complex processes. For instance, many of the problems regarding mathematical simulations in sociology, psychology, and politics actually stem from unknown order parameters.

At the same time, we know quite a lot about the order parameters of equation (4.2) governing distributed systems. In some cases they are

Fourier harmonic amplitudes, in others they are either the slowest or the fastest variables. Many important results have been obtained in this area. In some cases, in particular, investigators have been able to show and explain how a system composed of many mutually exclusive subsystems acquires features that do not exist in any of the subsystems.

Until quite recently, opinions differed as to whether isolation of order parameters was merely a heuristic procedure supported by considerations of interest or utility or whether results could be obtained in time through rigorous analysis. Among other things, a proof was much sought after that attractors for equations of form (4.2), which formally give systems with an infinite number of degrees of freedom, are described by the infinite-dimensional system (4.1). It seems that not long ago the majority of workers took the heuristic view.

The latest studies, in particular those of the American mathematicians Foias and Sell and the French mathematician Temam, show that rigorous results can be obtained for many equations, for example, for the series of reaction–diffusion systems [4.13]. The concept of order parameters has proved to be even more profound than it earlier seemed.

4.4 Strangeness of the Strange Attractors

When our results concerning the instability of non-periodic flow are applied to the atmosphere, which is ostensibly non-periodic, they indicate that prediction of the sufficiently distant future is impossible by any method, unless the present conditions are known exactly. In view of the inevitable inaccuracy and incompleteness of weather observations, precise very-long-range forecasting would seem to be non-existent.

Edward Lorenz

In discussing a significant scientific achievement or a novel idea, it is always interesting to see what the author's forebears and contemporaries thought about the matter. It often turns out that they expressed ideas that were much in accord with what eventually constituted an important discovery. The idea was actually 'in the air'. The discovery of chaos in determinisic systems was no exception.

Ray Bradbury gave a fine nutshell rendering of the idea in his story 'The Sound of Thunder'. A company uses a time machine to transport its clients back to the past so that they can go on safari for prehistoric animals. The company is extremely careful over picking animals for shooting, and the hunters have to use a special path so as not to affect the world of the past in any way. However, the hero of the story steps off the path by accident during an unfortunate safari and crushes a golden butterfly. Back in his own time, he realizes how dramatically the destruction of the butterfly has affected the course of events on Earth. The chemical composition of the air, tints of colors, even spelling rules are not the same as he had left them. Worst of all, the results of the presidential election were different and a new cruel regime has come to power. In the last moment of his life he realizes that a small thing like a golden butterfly had upset subtle balances and "knocked down a line of small dominoes, and then big dominoes and then gigantic dominoes".

Let me now quote some reflections on predictability expressed by Richard Feynman in one of his Lectures on Physics:

It is usually thought that this indeterminacy, that we cannot predict the future, is an important quantum-mechanical thing, and this is said to explain the behavior of the mind, feelings of free will, etc. But if the world were classical – if the laws of mechanics were classical – it is not quite obvious that the mind would not feel more or less the same. It is true classically that if we knew the position and the velocity of every particle in the world, or in a box of gas, we could predict what would happen. And therefore the classical world is deterministic. Suppose, however, that we have a finite accuracy and do not know *exactly* where just one atom is, say, to one part in a billion. Then as it goes along it hits another atom, and because we did not know the position better than to one part in a billion, we find an even larger error in the position after the collision. And that is amplified, of course, in the next collision, so that if we start with only a tiny error it rapidly magnifies to a very great uncertainty... Obviously, we cannot really predict the position of the drops unless we know the motion of the water *absolutely exactly*.

Speaking more precisely, given an arbitrary accuracy, no matter how precise, one can find a time long enough that we cannot make predictions valid for that long a time. Now the point is that this length of time is not very large... The time goes, in fact, only logarithmically with the error, and it turns out that in only a very, very tiny time we lose all our information [4.2].

The Feynman Lectures were published in the U.S.A. in 1963. In the same year, in the Journal of the Atmospheric Sciences, there came a paper by the American meteorologist Edward Lorenz initiating a new direction in natural history – that of investigation of chaos in deterministic systems [4.14].

One can only wonder at the boldness Lorenz showed in picking the simplest model, a system of only three ordinary differential equations, and after computing it through stating that what he was dealing with was a new phenomenon and not a computational error. Lorenz's findings can be illustrated by some graphic examples. Along with the attractors depicted in Figs. 4.1, 4.2, and 4.5, there exists one more type of attractor. It corresponds to processes in which the quantities under study never, even as $t \to \infty$, reach the steady state (as in Fig. 4.1), and never become either periodic (Fig. 4.2) or quasiperiodic (Fig. 4.5). No matter how long we watch them, they vary non-periodically, as shown, for example, in Fig. 4.13.

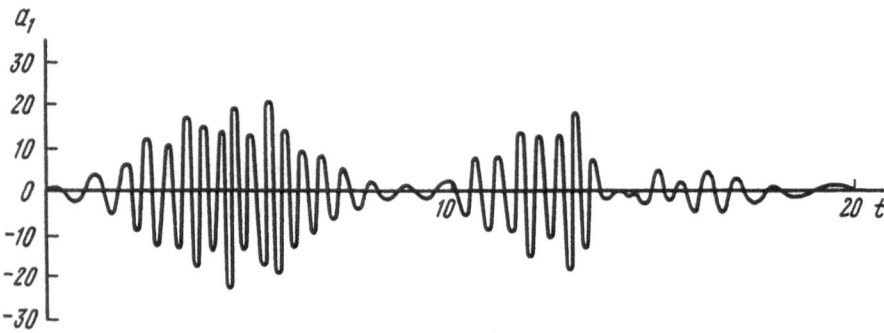

Fig. 4.13. Plot of some components typical of a non-periodic solution (in this case the curve indicates variation of one of the magnetic field components in the solar dynamo model; curves of this sort also indicate variations of solar activity)

The limit cycle has the appearance of a loop (see Fig. 4.2) that can be bent in a rather complicated manner in phase space. The invariant torus, on the other hand, looks like an infinitely long thread wound on a smooth surface. Unlike them, the strange attractor has a rather peculiar geometry and sometimes resembles a ball of trajectories, as shown in Fig. 4.14a. The attractor we see in the figure was first encountered when a problem in the theory of reaction–diffusion systems was investigated [4.10]. In that case, $N = 3$, as it did in Lorenz's system. Figure 4.14b shows the projection of this attractor onto a plane in three-dimensional phase space. It resembles two bands glued together. We can appreciate the remarkable geometry of strange attractors even more if we go back to the beginning

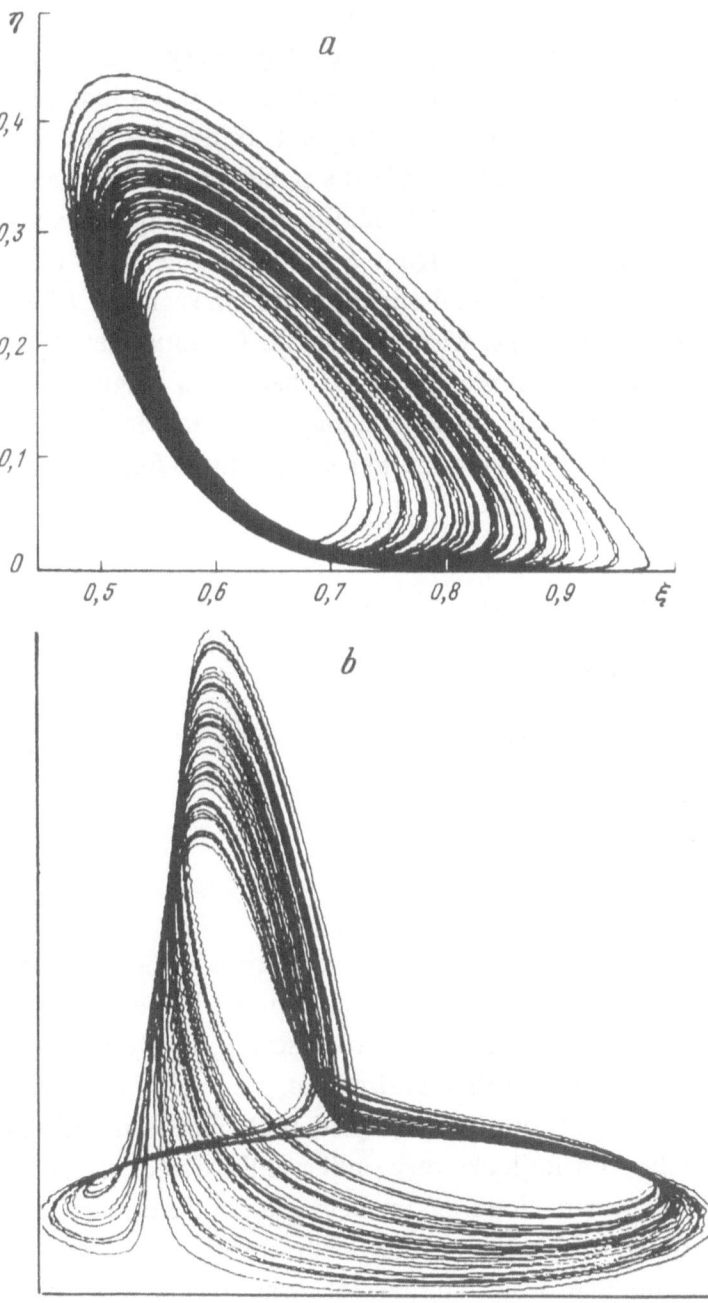

Fig. 4.14a,b. Projection onto a plane (**a**) and stereographic projection (**b**) of a strange attractor for a reaction–diffusion model

of this century, when problems of rigorous mathematical analysis were discussed just as widely as in more recent times.

At that time, one of the originators of contemporary mathematics, Georg Cantor, constructed a set C in which each point was a limit point (that is, any of its neighborhoods contained points of the set). The set C contains no intervals and is uncountable (its points cannot be counted with the natural numbers). The method for constructing the set is illustrated in Fig. 4.15: starting from a single segment, at every successive step we delete the central one-third of each segment. After a countable infinity of steps we get what is called the Cantorian set C. The set C possesses *scale invariance*. This means that at low magnification only two segments are visible. At higher magnification each segment splits into two. At still higher magnification, each of these four segments splits again into two, and so forth.

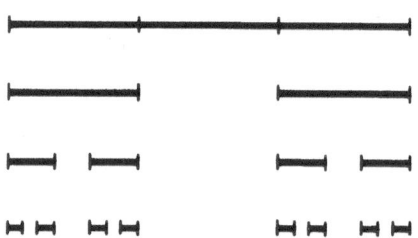

Fig. 4.15. Construction of the Cantorian set C

Now suppose that the bands of the attractor shown in Fig. 4.14b are intersected by some surface. We can see a set of points in the vicinity of the two continuous curves (Fig. 4.16a). After scaling up we see more points (Fig. 4.16b). After further scaling up, i.e., with increasing resolution, the band splits into two (Fig. 4.16c). One can see that sectional view of the attractor looks very similar to the Cantorian set. An unusual geometrical construction from the beginning of the century turns out to be typical of the whole class of strange attractors.

It is often said that a self-similar set, when scaled up, has a *Cantorian structure* like that of the set C. Analysis of these sets indicates that the usual measures of length, area, and volume are inefficient in their case. Indeed, if we calculate the length of a Cantorian set by going to the limit we find that it is zero. Even if we delete one p-th of each segment instead

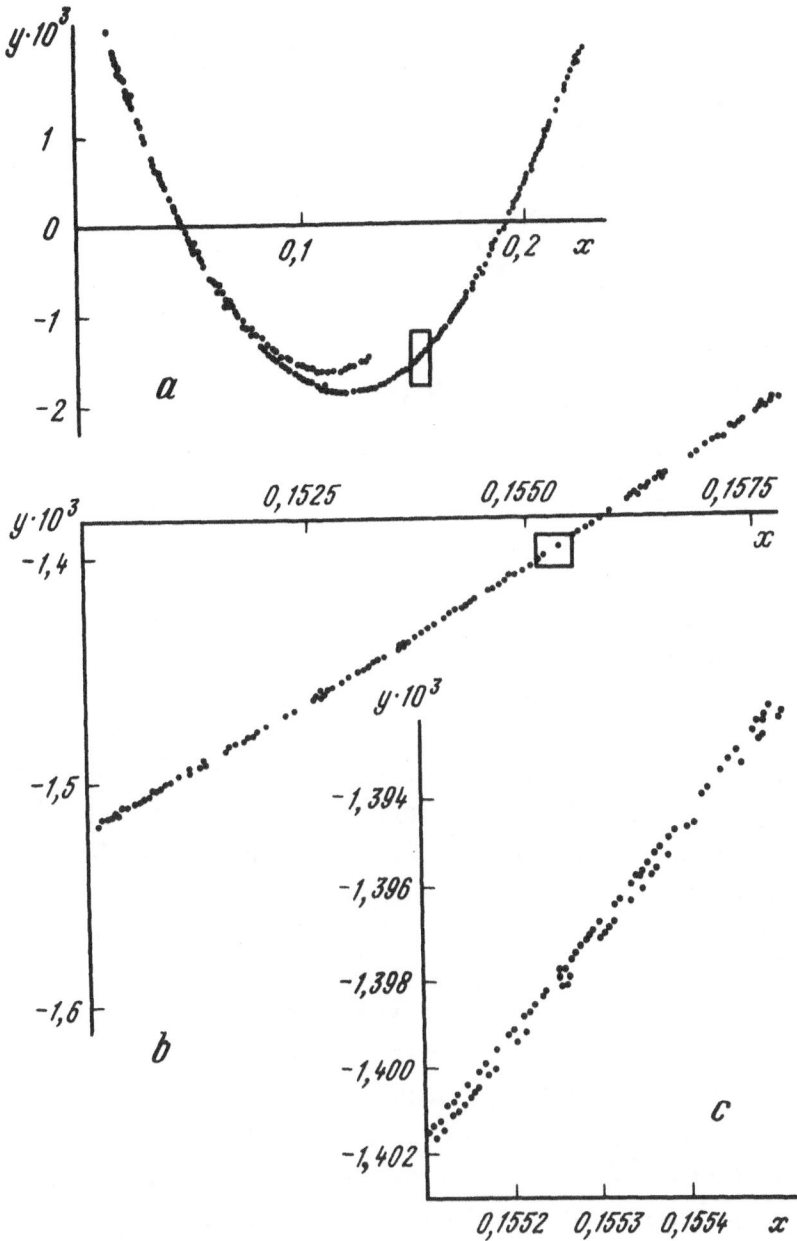

Fig. 4.16. Cantorian structure of the attractor shown in Fig. 4.14

of one-third, the length will still equal zero. However, the resulting set in this case will be different. One can easily model the situation by deleting smaller and smaller parts of the remainder until the resultant set has finite length, which is all the more paradoxical since the set still contains no intervals, as before.

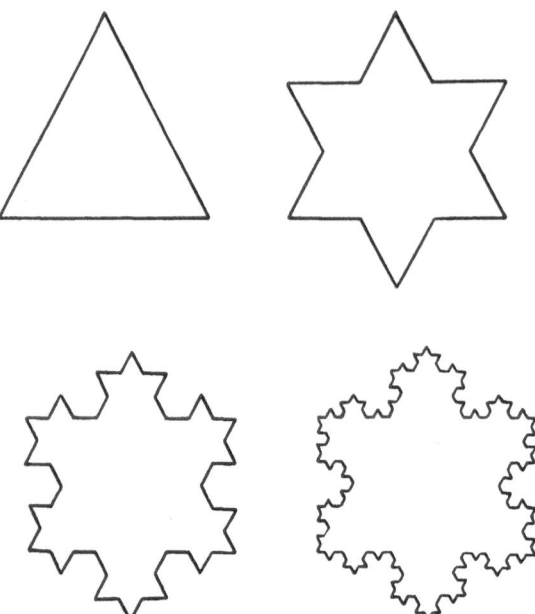

Fig. 4.17. Construction of Koch's island

There is one more mathematical phenomenon worthy of note, namely, Koch's island. Its construction is shown in Fig. 4.17. By inspection, one can see that the area of the figure obtained in the limit (Koch's island itself) is bounded, whereas its perimeter grows without bound. Suppose that we want to measure the perimeter using a ruler with scale units of size ε. It will be obvious from the construction scheme that the measured perimeter length L depends on ε: $L(\varepsilon) \approx C\varepsilon^{-\alpha}$, $\alpha > 0$. It is only natural to assume that α is the most substantial factor when we are analyzing objects of Cantorian structure.

Quantities similar to α were introduced in 1919 by Felix Hausdorff. For the simplest case of a 'good' set the underlying idea is this. Assume that within a p-dimensional space we cover the set with p-dimensional balls of radius ε. The number of balls is different for different coverings. Take a covering requiring a minimal number of balls and denote it by $N(\varepsilon)$. Assume that as $\varepsilon \to 0$, $N \approx \varepsilon^{-d_c}$, where d_c is the *capacity* of the set, and one of its important characteristics [4.15].

To obtain a minimal covering for a segment in a one-dimensional space ($p = 1$), one breaks down the segment into lengths ε and has $N \approx \varepsilon^{-1}$ and $d_c = 1$. For a square, $d_c = 2$, and for a cube, $d_c = 3$. Quantity d_c here actually coincides with dimension.

For a Cantorian set C, it seems reasonable to proceed as follows. Take $\varepsilon = \frac{1}{3}$. To cover C, one needs two such lengths. If $\varepsilon = \frac{1}{9}$, four lengths are required, if $\varepsilon = (\frac{1}{3})^n$, one needs 2^n lengths, and hence $d_c = \ln 2/\ln 3$.

Now, if we delete one s-th instead of one-third of each segment, $d_c = \ln 2/\ln s$. It appears that d_c is better than ordinary length for representing Cantorian structure. Sets with fractional capacities (or their analogues) are referred to as *fractals*. Most strange attractors fall into this category.

Knowing the fractal dimension is very helpful for making forecasts. Suppose that an object under discussion is given by a system of equations of form (4.1) with N as the most essential characteristic. Assume that we have an underestimate M of N (that is, we consider the system to be simpler than it actually is). Then, on the one hand, we shall not measure all the existing order parameters, and two different states of the system will be taken for one state. On the other hand, the system will be subject to unexpected evolution from time to time.

By way of example, consider some fanciful two-dimensional creatures living on a sphere ($M = 2$). They may not suspect the existence of the third dimension ($N = 3$), and therefore the appearance of an object out of the third dimension onto the sphere will be regarded as a miracle (see Fig. 4.18). The same might happen to us if we underestimate N. It is probably our ignorance of N that restricts utilization of dynamical systems of form (4.1) in non-traditional fields of modeling such as sociology, psychology, politics, and so on.

Fig. 4.18. A star coming from the third dimension as perceived by 'planarians' living on a sphere

Simply put, the connection between N and d_c is this: if an attractor belonging to a real process has dimension d_c, then in its model of the form (4.1), $N \geq d_c$ in the case that d_c is an integer, and $N \geq [d_c]+1$ otherwise (where $[d_c]$ is the integral part of d_c). Actually, however, the connection is far more profound, and stems from another fundamental property of strange attractors – their *sensitive dependence on initial data*.

Strange attractors are characterized by at least one positive Lyapunov exponent λ_1. An interesting point about positive Lyapunov exponents is that they enable an approach to the forecasting of natural phenomena. Indeed, when the attractor is a cycle or a point, it is possible through lengthy observation to make a good forecast even if a small error has crept into the initial knowledge of $x(t)$, since $|x(t) - x'(t)|$ does not grow. However, in time $\tau \approx 1/\lambda_1$ two originally close strange attractor trajectories are no longer close. Regardless of how small the error may be, in principle we cannot predict what will happen to the system in time τ.

The system is predictable for much shorter times. Statistical properties like the mean value of $x(t)$, variance, and so on, can be determined by tracking the system's behavior over sufficiently long times. The scope for forecasting $x(t)$ is nonetheless limited.

Lorenz attributed the problems involved in medium-term (several weeks ahead) forecasting to this sensitivity to initial conditions. In fact, neither improved mathematical models, nor powerful computers, nor novel numerical methods can ensure efficient medium-term forecasting. When we recall that the atmosphere is modeled by some dynamical system of the form (4.1) and that its highest-order Lyapunov exponent $\lambda^{-1} \approx 1$ week, we can perhaps realize the difficulties encountered by meteorologists.

An N-dimensional dissipative dynamical system described by an equation of the form (4.1) generally has N Lyapunov exponents depending on a given trajectory $x(t)$ on an attractor. The Lyapunov exponents are the same for almost all trajectories. For a strange attractor, the Lyapunov exponents may include a certain number of positive $\lambda_1,...,\lambda_k > 0$ corresponding to close trajectories diverging from a given one, one zero $\lambda_{k+1} = 0$, and $N-k-1$ negative $\lambda_{k+2},...,\lambda_N < 0$ giving closely spaced trajectories tending to $x(t)$.

A fascinating problem is the connection between dynamical characteristics (e.g., λ_i, $i = 1,...,N$) and geometrical ones (e.g., d_c). The

American mathematicians Caplan and Yorke proposed a hypothesis implying that

$$d_c = j + \frac{\sum_{i=1}^{j} \lambda_i}{\left|\lambda_{j+1}\right|} \, ,$$

where j is the greatest value such that

$$\sum_{i=1}^{j} \lambda_i > 0$$

[4.16]. This formula has been rigorously validated for a large variety of continuous-time systems of the form (4.1) with $N = 3$ and for discrete-time ($x_{n+1} = g(x_n)$) systems with $N = 2$. At larger N, the situation is more complex and is now an active research area, but even there the Caplan–Yorke formula provides a good approximation.

However, the close connection between the geometry and the dynamics of strange attractors is striking. By measuring the Lyapunov exponents of a strange attractor one can form a judgement about its geometry, while knowledge of its capacity gives information about its Lyapunov exponents.

Another important question is the following. To what extent is the behavior of a system containing a strange attractor random? Let us return to Fig. 4.14b. When a display screen is used to show a point representing a system's state traveling along its attractor, one can see that the point randomly falls now on the left, now on the right band. Here it seems reasonable to represent such a trajectory as an infinite series of binary digits. If at the k-th turn the point runs along the left band, we write 0 for the k-th point of the series, and if it runs along the right band we write 1. Thus we obtain an infinite series of 0s and 1s (which is specific for each individual starting point x_0).

Coming now to the theory of probability, consider the trivial example of flipping a coin. Suppose we flip a coin an infinite number of times. The results can be written as a series $\{b_k | k = 1, \ldots\}$. If at the k-th flip we get heads, we write 0 for the k-th term of the series; if tails, we write 1. What is important is that the series $\{b_k\}$ defines a certain number in the binary system. Each possible infinite series of 0s and 1s corresponds to a particular real number. (There are uncountably many non-periodic sequences among all the possible series $\{b_k\}$. It is interesting to note that a one-to-one correspondence exists between the uncountable infinity of infinite series of 0s and 1s and the infinity of points in a Cantorian set.)

It has been proved that it is possible to find a starting point x_0 which represents the series $\{c_k\}$ corresponding to any chosen coin-flipping series $\{b_k\}$.

This is surprising. Indeed, imagine there is an instrument capable of measuring only the sign of x. Then, irrespective of the number of measurements we make, it is impossible to determine whether the process is purely random, as in the coin-flipping case, or deterministic, as in model (4.1).

Using systems containing strange attractors we can apparently model various phenomena, such as oscillatory chemical reactions, fluctuation of species populations, fluid turbulence, or certain economic processes. We come across chaos everywhere, even in the simplest systems. Let us consider two examples.

Solar activity shows up locally in the forms of sunspots, eruption of prominences, and so on, with lifetimes never exceeding several revolutions of the Sun. But the activity is not random. It is modulated by a mechanism with both large-scale and long-term characteristics. The most essential of these is a 22-year recurrence period for the magnetic field and the associated 11-year cycle called Wolf's variation in the number of sunspots, $W(t)$. Apart from this, an appreciable modulation with an approximately 57-year period takes place on the Sun. There are also global attenuations of activity occurring in cycles exceeding several decades. The best known of these is the so-called Maunder minimum. The long-term modulation of solar activity is easily tracked by measuring concentrations of the carbon-14 isotope in the annual growth rings of trees. The modulation is chaotic and thus suggests the existence of a strange attractor in the dynamical system governing solar activity.

The manner in which the attractor is projected onto models of the form (4.1) of long-term variation of solar activity is shown in Fig. 4.19, where $N = 7$. The first four equations give the Sun's magnetic field, and the remaining three its hydrodynamic motion. The dimensionless parameter D is proportional to the spin velocity. At comparatively small D, the attractor may become an invariant torus (Fig. 4.19a). At large D it is strange, though lying in the torus neighborhood (Fig. 4.19b). The 57-year periodicity modulation is probably related to this fact.

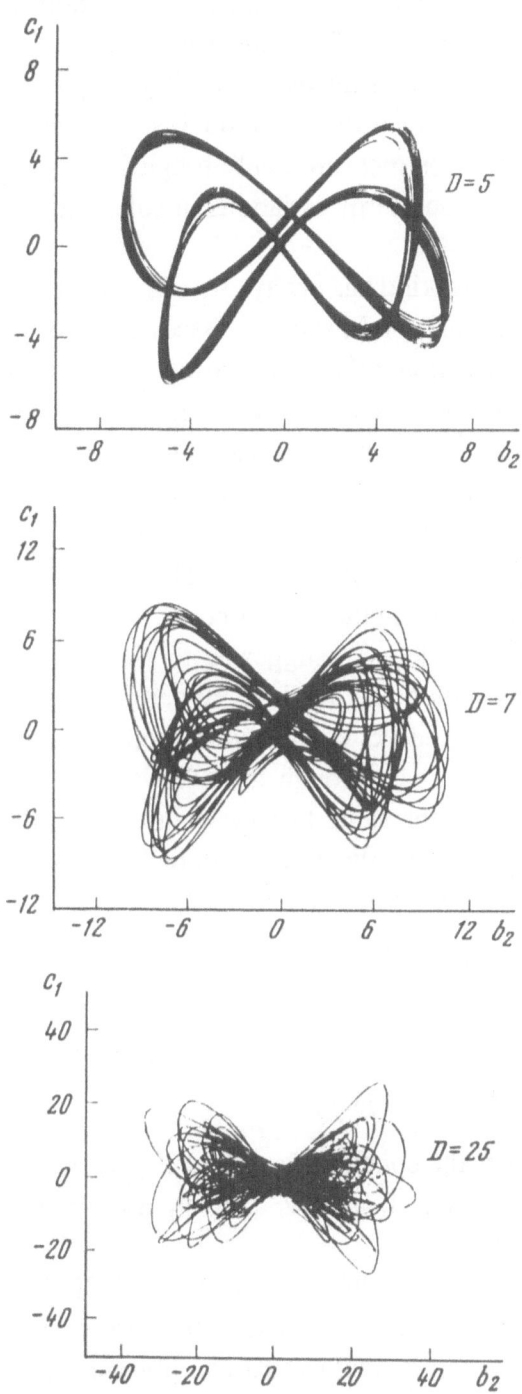

Fig. 4.19. Typical appearance of an attractor in the solar dynamo model

Then, however, the attractor's capacity d_c goes up, and the motion on it becomes more complicated (Fig. 4.19c). This model is in qualitative agreement with observational results and makes it possible to predict effects like the Maunder minimum. Given that the model is sensitive to initial conditions and that $W(t)$ can be measured only roughly, one can hope to forecast only some of the long-term variations in solar activity [4.17].

Let us take another example. Mankind has always striven to look into both the remote future and the distant past. In some instances this appears possible. Thus, knowing the direction of magnetization of old rock samples one can determine the positions of the Earth's magnetic poles as they were millions of years ago, as well as their displacement. The results reveal that over perhaps the last 600 million years the position of the magnetic poles has varied chaotically. It is extremely important to understand the mechanisms underlying such behavior. They will probably enable us to look into what goes on in the Earth's core.

In 1958 an original physical model was suggested to simulate the poles' chaotic motion. This is called the Rikitaki dynamo after its originator. It consists of two dynamo disks connected in such a manner that current coming from one disk goes through the other's coil (Fig. 4.20). The shafts are acted upon by the same torque. In this model the disks represent two gigantic whirls in the Earth's core [4.18]. The governing equations are really very simple:

$$\dot{x}_1 = -\mu x_1 + x_2 x_3$$
$$\dot{x}_2 = -\mu x_2 + x_1 x_4$$
$$\dot{x}_3 = 1 - x_1 x_2 - v_1 x_3$$
$$\dot{x}_4 = 1 - x_1 x_2 - v_2 x_4 ,$$

where x_1 and x_2 are currents, μ is the conductor's resistance, x_3 and x_4 are the angular velocities of the disks, and v_1 and v_2 are friction coefficients.

Strictly speaking, these equations produce the generalized Rikitaki model that accounts for the shafts' friction. The model seems to be quite good at representing the true process. It implies that equilibrium states, i.e., stability points, can be found at any parameter values. Furthermore, there may exist limit cycles and strange attractors. In other words, under certain initial conditions $x_1(0)$, $x_2(0)$, $x_3(0)$, $x_4(0)$ the polarity of the

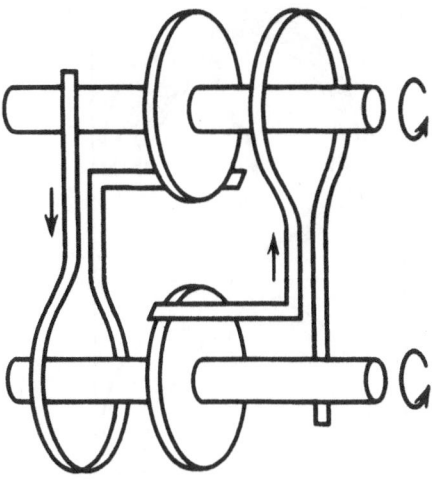

Fig. 4.20. Schematic diagram of the Rikitaki dynamo

magnetic field remains the same, while under other conditions it varies periodically or chaotically [4.19].

One of the attractor's projections obtained at different values of μ is shown in Fig. 4.21. As one can see, it has a rather complicated structure including both limit cycles and strange attractors. In some cases they are symmetrical, in others they are not.

Each of the strange attractors is a complex object, and its detailed analysis is far from simple even with modern mathematics. Presuming that one of the system's parameters (say, μ) varies very slowly (at a rate mucher slower than that of reaching the attractor), the attractor's evolution with changing values of the parameter becomes a matter for prediction.

How can this problem be approached? Proceeding from the results discussed in Sect. 4.2, the best approach appears to be as follows. First, using measured values of the order parameters and simplified models, the transition from simple attractors to more complex chaotic is analyzed. Then the sequence of bifurcations that lead to an increase in the complexity of the attractor is clarified.

The analysis of mechanisms underlying the transition from order to chaos under changing parameters in various models and real systems is now often referred to as the analysis of transition-to-chaos scenarios. The results obtained in recent years suggest that only a few universal scenarios can in general be implemented in nature. The discovery of this

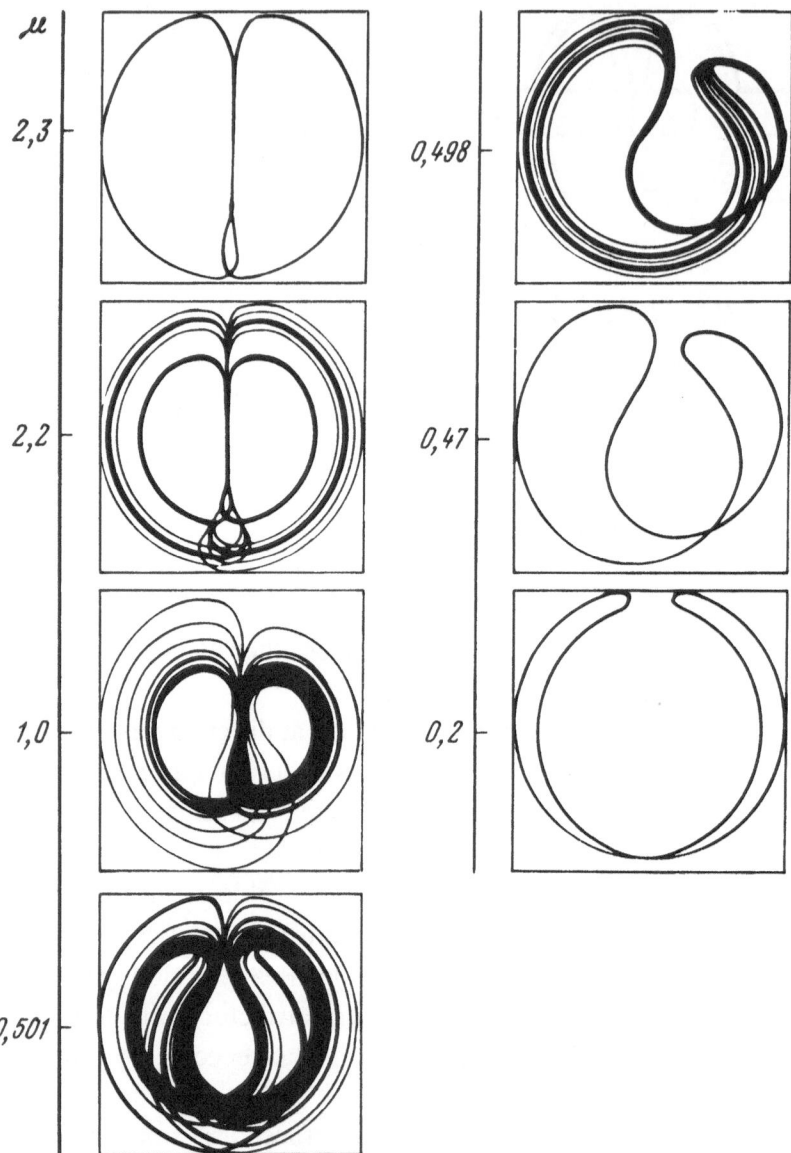

Fig. 4.21. Attractor evolution in the generalized Rikitaki model at varying parameters ($\nu_1 = 0.004$, $\nu_2 = 0.002$)

surprising fact is perhaps one of the most important findings of applied mathematics made in this century. Among other things, it has enabled us to achieve a better understanding of the inherent unity of nonlinear phenomena.

An object that is strikingly simple on a superficial view, the so-called one-dimensional mapping, can be used to model the situation:

$$x_{n+1} = f(x_n, \lambda).\tag{4.3}$$

If $f(x, \lambda) = x + \lambda$, then the sequence $\{x_n\}$ of numbers is an arithmetic progression. If $f(x, \lambda) = \lambda x$, then $\{x_n\}$ is a geometric progression. However, when f is a nonlinear function, no matter how simple it might be, e.g., the quadratic parabola $f(x, \lambda) = \lambda x(1 - x)$, the above one-dimensional mapping turns out to be very interesting. The majority of problems connected with this model could have been stated as far back as in Euclid's time. Their investigation was not possible in those days, however, since it requires extensive computational experiments and a quite new body of mathematics. In fact, it was actively embarked upon only in the 1970s [4.20].

The study of one-dimensional mappings in which $f(x, \lambda) = \lambda x(1 - x)$ made it possible to find two universal scenarios for the transition from order to chaos in dynamical systems. It is precisely these scenarios that were found with time to be implemented in many hydrodynamic, chemical, and radio-electronic systems, and our present understanding of them is very helpful for making predictions. The fundamental significance of these results for synergetics and nonlinear dynamics has been discussed many times.

There is another and lesser-known aspect of the matter. The analysis of one-dimensional mappings reveals a situation we may encounter when making predictions for real systems. Suppose that mapping (4.3) has generated sequence $\{x_n\}$ belonging to an isolated line segment. Its behavior as $n \to \infty$ is determined by an attracting set – an attractor.

To describe these attractors, we again introduce the Lyapunov exponent μ. Since the mapping is one-dimensional, we use only one exponent. At $\mu < 0$, the system is predictable. At $\mu > 0$, the above mentioned sensitivity to initial data comes into play once more.

The following simple technique can be used to give an idea of the one-dimensional mapping's attractor. Let us divide the segment into M smaller segments of length ε ($\varepsilon = 1/M$) and see which segments the first N elements of the sequence $\{x_n\}$ fall into. Plotting the numbers of elements on the segments yields step diagrams, or histograms (see, e.g., Fig. 4.22a). Passing to the limit $M \to \infty$, $N \to \infty$, $\varepsilon \to 0$ and

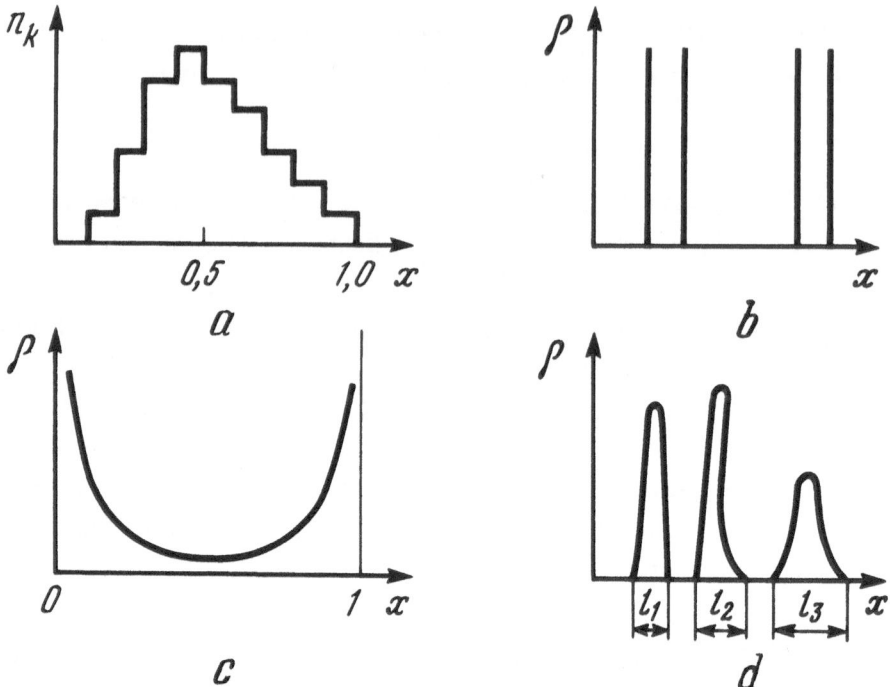

Fig. 4.22. (a) Example of a histogram for $M = 10$, $\varepsilon = 1/10$, $N = 30$, (b–d) invariant measures of one-dimensional mappings

normalizing the area enclosed by the diagram to unity, we get the dependence of the probability density p on x.

Now, exactly what $p(x)$ functions will our mapping's attractors have? The simplest case is that of $\{x_n\}$ tending to a periodic sequence $\{a_n\}$ with $a_n = a_{n+r+1}$. This is actually a limit-cycle analogue of a periodic continuous system referred to as a stable cycle. The function $p(x)$ in this event shows up as r identical peaks (see Fig. 4.22b). (It is acceptable to say that each of the peaks has infinite height, zero width, and finite area $1/r$). The Lyapunov exponent is negative, and long-term prediction is possible.

Another case is realized at $\lambda = 4$. Elements of the sequence $\{x_n\}$ fill in the entire segment, though at varying density (see Fig. 4.22c). Now the Lyapunov exponent is positive and it is impossible to make a long-term prediction. This is the analogue of the strange attractor for the mapping. In fact, such attractors are not infrequent in practice and do not seem peculiar.

Yet another, quite different case is realized at certain values of λ. Its typical $p(x)$ is shown in Fig. 4.22d. Here $p(x) \neq 0$ only within several 'islands' (three in the case shown). All elements x_n such that $n = 3k$, $k = 1,2,...$ belong to the first island, those with $n = 3k + 1$ to the second island, and those with $n = 3k + 2$ to the third. In other words, the order in which elements are assigned to the respective islands is strictly fixed.

In this case, the Lyapunov exponent is strictly positive, and there seems to be no hope of making a global prediction. However, given an accuracy of $L > l \equiv \max(l_1, l_2, l_3)$ (see Fig. 4.22d), one can make such a prediction due to the fixed order of visiting the islands. The attractors involved are known as 'noisy' cycles. If such cycles occur in many real systems, then new opportunities may exist to solve prediction problems for those systems.

Another feature of nonlinear systems revealed by one-dimensional mapping has to do with their dependence upon parameters. In the mapping (4.3) involving function $f(x, \lambda) = \lambda x(1 - x)$, the range of parameter variation coincides with the line segment $0 \leq \lambda \leq 4$. Roughly speaking, we can find out whether order or chaos occurs at each specific parameter value.

It is natural to suppose that if chaos occurs at $\lambda = \lambda_0$, then it also occurs at nearby parameter values $(\lambda_0 - \delta) < \lambda < (\lambda_0 + \delta)$. If this were the case, the system could be called stable with respect to changing λ. But the situation is actually quite different: in arbitrarily close neighborhoods of point λ_0 there exist values of λ corresponding to stable cycles and hence to order.

However, given an arbitrary value of λ in the range $0 \leq \lambda \leq 4$, the system has a high chance of behaving chaotically. This paradoxical assertion was proved by the Soviet mathematician M. Jakobson. His American colleague Farmer described the phenomenon itself as sensitive dependence on parameters [4.21, 4.22].

Imagine that a system under study is precisely represented by a known one-dimensional map and its initial condition x_1 is given accurately. If we do not know the value of λ exactly, even if the uncertainty is as small as desired, the prediction may become merely probable again. With some probability the system is chaotic.

4.5 Dynamical Chaos and Reality

This you must know!
From one make ten,
And two let go,
Take three again,
Then you'll be rich.
The four you fix.
From five and six,
Thus says the witch,
Make seven and eight,
That does the trick;
And nine is one,
And ten is none.
That is the witch's arithmetic.

Goethe's Faust
translated by W. Kaufmann

Let us visualize the following experiment. Let a phenomenon under study be given ideally by a differential equation involving a strange attractor. How can theory be confronted with experimental results in this case? Assume that the system is sensitive to initial conditions and that two close trajectories rapidly diverge from one another. Then we can expect the difference between the computed trajectory and experimental data to grow with time. This is due not to the model's shortcomings but rather to the nature of the underlying dynamics.

Testing physical theories often involves comparing either a small number of measured quantities (frequencies, lengths, times, etc.) or quantities averaged over a large time interval. However, there is a large class of problems where we not only seek specific numerical values at a given point at a given time, or values averaged over a large time interval, but also want to know what the process dynamics itself is like. Among such problems are those concerned with predicting the behavior of nonlinear systems, for example, weather forecasting.

Whether a model applied to such problems is good or not can be judged by the accuracy of its predictions and the time resolution it provides, rather than by its success in averaging over large times.

In effect, the discovery of strange attractors has led to a drastic reformulation of the problem of agreement between experimental data and theory. How can one example of chaotic behavior be compared with another?

It has been believed for many years that the comparison should be made between certain quantities widely used in the theory of probability, such as mean values, variance, and probability densities $p(x)$. More recently, however, it has been shown that the problem is not as simple as it formerly seemed to be. A one-dimensional mapping $x_{n+1} = f(x_n)$ was shown to have an infinite number of different functions f (governing the dynamics) corresponding to one and the same probability density $p(x)$ (a statistical characteristic). In order that this correspondence be one-to-one, it is necessary not only to specify the number of $p(x)$ peaks, but to give some nontrivial information about the symmetry of f as well [4.23].

Another approach, currently thought to be the best for this application, consists in comparing the model attractor's characteristics (capacity, Lyapunov exponents, etc.) with those of an experimentally described one. This procedure is somewhat obscure and needs to be explained.

Returning to the experiment we staged mentally at the beginning of this section, let us discuss what exactly we should measure, in what way, and what characteristics we should derive thereby, in order to investigate chaotic behavior in nonlinear media. Suppose we have an instrument that measures one of the system's characteristics at different times spaced at intervals Δt. The measurement results yield a bounded sequence $\{a_i\}$, $0 \leq i < \infty$. The following question now arises: can we use this system to determine whether we are dealing with a complex deterministic process governed by the differential equation $dx/dt = f$ in an n-dimensional phase space or with a random function? The answer was given in 1980 by Takens, presently one of the best-known experts on chaotic systems [4.24]. It was he and his colleague Ruelle who introduced the term 'strange attractor' in 1971.

Presumably, the measurement results can be *described using a smooth deterministic model* if there is a differential equation with a smooth function $f(x)$ and a smooth function g such that

1) for each observable sequence $\{a_i\}$ of experimental data, there is a point x_0 such that $a_i = g[x(i\Delta t)]$, where $x(t)$ is a solution to the differential equation with initial conditions $x(0) = x_0$;

2) for each starting point x_0, the solution $x(t)$, $x(0) = x_0$ is bounded at $t > 0$.

If functions $\mathbf{f}(\mathbf{x})$ and g as well as $\mathbf{x}(t)$ can be obtained for some sequence $\{a_i\}$, the results can presumably be explained using a smooth deterministic model.

In order to formulate the Takens criterion, we need to introduce several notions. Let $\{a_i\}$ be a bounded sequence of real values (experimental data). We determine the set of positive integers $\underline{C}_{n,\varepsilon}$ for each $\varepsilon > 0$ and positive integer n in the following way: $0 \in \underline{C}_{n,\varepsilon}$, and at $i > 0$, $i \in \underline{C}_{n,\varepsilon}$ if and only if for all $j < i$, $j \in \underline{C}_{n,\varepsilon}$,

$$\max\left\{\left|a_i - a_j\right|, \left|a_{i+1} - a_{j+1}\right|, \ldots, \left|a_{i+n} - a_{j+n}\right|\right\} \geq \varepsilon.$$

Denote the number of elements in $\underline{C}_{n,\varepsilon}$ as $\underline{C}_{n,\varepsilon}(\{a_i\})$. Since $\{a_i\}$ is a bounded sequence, $\underline{C}_{n,\varepsilon}$ is a finite quantity.

Takens states the following: experimental results defining sequence $\{a_i\}$ can be explained by a smooth deterministic model if quantity $\ln \underline{C}_{n,\varepsilon}(\{a_i\})/(n - \ln \varepsilon)$ is uniformly bounded as $(n - \ln \varepsilon) \to \infty$. If the sequence is unbounded, a smooth deterministic model cannot account for the sequence of measurements [4.24].

In fact, the last statement makes it possible to distinguish between chaos in a deterministic system and randomness, and to see whether there is any order within the chaotic system under investigation.

The Takens theorem, the field of its application, and a number of associated details will be taken up separately. It should be borne in mind, however, that an abyss exists between the theory and experiment, even in the ideal case where we deal with an infinitely long sequence of accurately measured values. We find a limit in the Takens theorem at $\varepsilon \to 0$. It is only natural to expect that a direct computation using a computer would continue infinitely and never stop.

In the 1980s a bridge was thrown over the abyss in the form of computational algorithms for analyzing experimental data on chaotic systems. The strength and reliability of this bridge determines whether some elegant results on chaotic dynamics are viable or not.

There is a notion often reported in popular science books and stemming from one of the originators of quantum mechanics about the role of 'crazy ideas' in the development of new knowledge. In the fifty years that have passed since this phrase was first used, the world has of course changed. The development of formalism, the availability of computers, and new experimental techniques are now the main factors enabling a deeper comprehension of the laws of nature.

The importance of 'craftsmanship' in research work is now growing rapidly. It suffices to look through the list of Nobel prize winners to see that it is an excellent performance rather than a brilliant idea that most often receives recognition. Apparently, the concepts of synergetics and chaotic dynamics have not yet found a sufficiently 'craftsmanlike' representation to turn contemporary research work into a routine business.

Returning to the bridge, let us direct our attention to several basic ideas and algorithms and afterwards discuss problems and difficulties.

The previously mentioned capacity d_c of a set is naturally insufficient to describe fully the 'ball of trajectories', and it cannot be estimated experimentally. That is why strange attractors are now usually characterized through an infinite number of *fractal dimensions*.

As a matter of fact, points defining a strange attractor occur in different regions of a phase space with different probabilities. To accommodate this fact, we use a cube of side ε to cover the test set in a p-dimensional space. A correlation is established between the i-th cube and the probability p_i that the set points fall into it. Now we introduce a set of quantities D_q, the so-called *generalized dimensions*, such that

$$D_q = \frac{1}{q-1} \lim_{\varepsilon \to 0} \frac{\log \sum_i P_i^q}{\log \varepsilon} \quad ,$$

where the summation extends over all covering cubes. D_q values are usually defined at all real q, and it is possible to say that $D_q > D_{q'}$ if $q' \geq q$. Such a set of generalized dimensions enables a more adequate description of the set under study than the dimensions taken separately.

The generalized dimensions were introduced in 1985 for a certain class of 'good' sets. At specific q values they coincide with the known dimensions. Thus, at $q = 0$ we are dealing with a capacity, since the sum in the numerator coincides with the number of cubes covering the test set. D_1 is called the *informational dimension* and D_2 the *correlation exponent* (commonly designated by letter v).

It is possible to determine the entire set D_q, $-\infty < q < \infty$, or, otherwise, to compute the α-spectrum, for several attractors to ensure that d_c and v do not differ much from other dimensions and quantities calculated using the Takens theorem (see Fig. 4.23). Therefore, we can estimate the quantity D_q that is most accessible to machine computation.

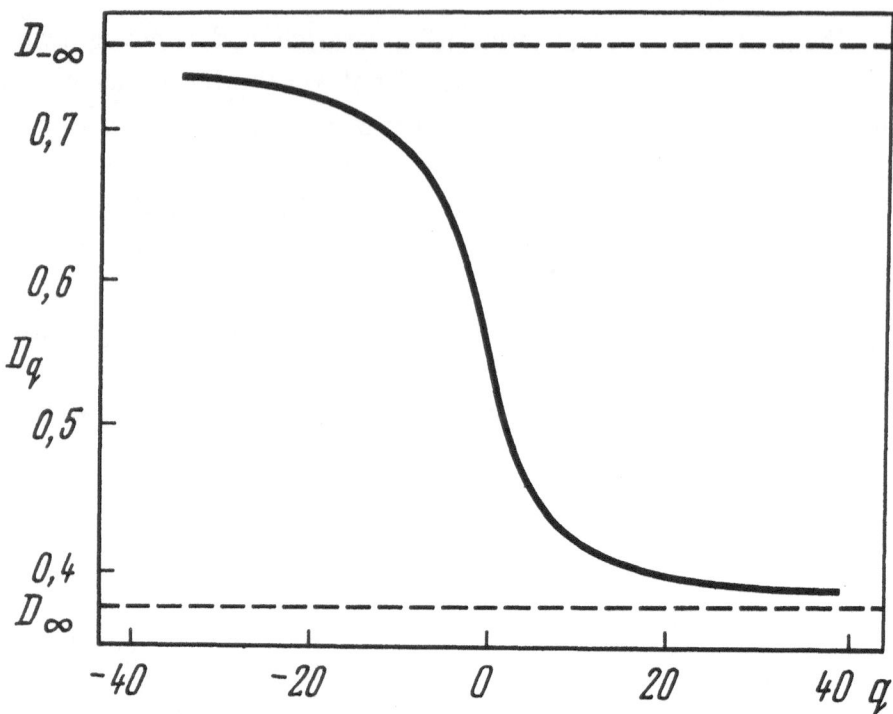

Fig. 4.23. Set of generalized dimensions for Feigenbaum's attractor. A switch from cycles to non-periodic trajectories occurs in the mapping $x_{n+1} = \lambda x_n (1 - x_n)$ as $\lambda \to \lambda_\infty = 3.5699456$. Feigenbaum's attractor is one that exists at $\lambda = \lambda_\infty$

The best quantity for this purpose is at present the correlation exponent. It was introduced in 1983 by Grassberger and Procaccia, who also proposed a technique to estimate it. Arbitrarily taking N points of the set under study in a p-dimensional space, we estimate the separations between all points, $p(\mathbf{x}_i, \mathbf{x}_j)$, and then functions

$$C(\varepsilon) = \lim_{N \to \infty} N^{-2} \quad \text{(the number of separations } p < \varepsilon\text{)}.$$

The correlation exponent is given by

$$v = \lim_{\varepsilon \to 0} \left[\frac{\ln C(\varepsilon)}{\ln \varepsilon} \right].$$

The following procedure is used to estimate an attractor's fractal dimension.

To acquire the experimental data needed to analyze a process taking place in a nonlinear medium, we measure some of its time-dependent characteristics a_i at one point (if a_i is a vector, we take any of its components) at fixed time intervals Δt. Using these data, we construct a set of m-dimensional vectors $\zeta_k = (a_k, a_{k+1}, ..., a_{k+m-1})$ (for example, $\zeta_1 = (a_1, a_2, ..., a_m)$). These are called ζ-vectors, and the embedding space is called ζ-space. Then we estimate the fractal 'dimension' (commonly, the correlation exponent) of the set of ζ-vectors. (The quotes are used because the dimension, i.e., the limit as $\varepsilon \to 0$, of any set composed of a finite number of points is zero. However, it is not the limit that we wish to know in practice, but rather the slope of the plot of $\ln C(\varepsilon)$ against $\ln \varepsilon$ within some range of values of ε.) We perform this procedure for $m = 1, 2, ...$ If an attractor has a finite dimension $\bar{p}$, we can expect that at fairly large $m > m_0$ (usually at $m > 2\bar{p} + 1$) the values obtained will be independent of m.

This suggests the existence of a set of m_0 order parameters to which all the rest system's degrees of freedom 'adhere'. Such behavior is characteristic of many mathematical models investigated in synergetics, including nonlinear systems with an infinite number of degrees of freedom given by partial differential equations.

The construction of ζ-vectors and estimation of $\bar{p}$ are referred to as *attractor reconstruction*. Among other things, it is employed in cases that go beyond those it was designed for. This gives rise to a number of interesting mathematical problems.

We note a paradoxical feature of this procedure. Measurement of just one quantity at discrete times suffices to analyze the number of order parameters for a complex multi-dimensional system.

The dimension of a strange attractor gives an idea of the attractor's geometry and the number of order parameters, though it offers no means to assess the rate of divergence of close trajectories. To make such an assessment possible, one needs to be able to compute Lyapunov exponents using experimental data.

The simplest and most helpful algorithm for this application was developed in 1985 and is called (conditionally) the analogue method [4.27]. Its basic idea is clear from Fig. 4.24a (compare this with Figs. 4.3, 4.4). We choose vector ζ_i (it can be a ζ-vector taken at a certain instant of time) and look for a close neighbor ζ_i' spaced from ζ_i by $L(t_0)$ at time t_0. If the difference $i - i_0$ is really large, it can be said that we are dealing with two closely separated trajectories. Since there is a positive index in the system, the separation between ζ_i' and ζ_i grows, reaching $L'(t_1)$ at instant t_1, by which time it can no longer be considered small.

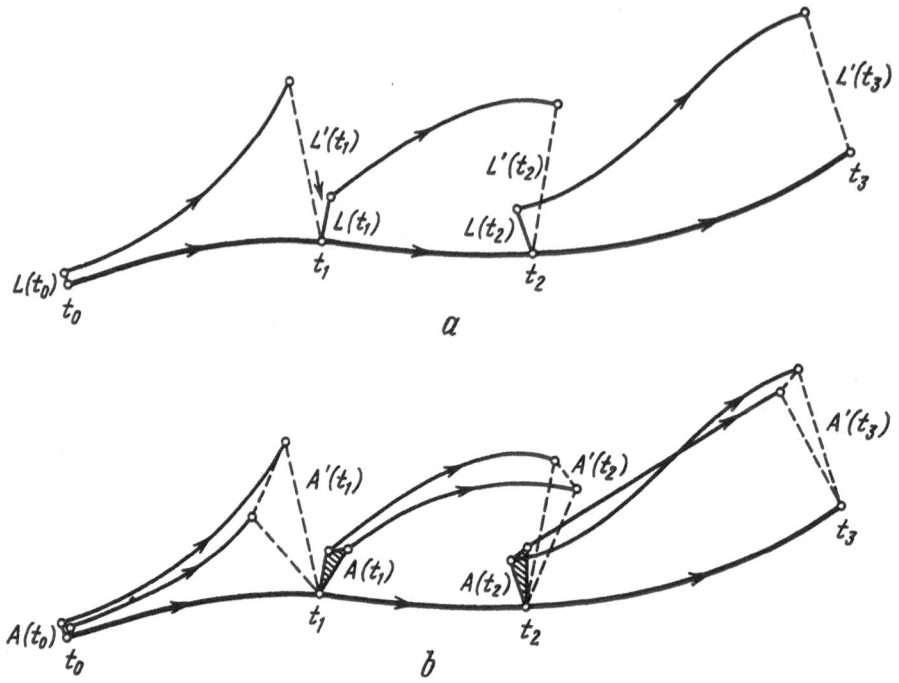

Fig. 4.24. Estimation of a Lyapunov exponent using the analogue method

Another close neighbor ζ_i'' is chosen at distance $L(t_1)$ so as to obtain a close parallel between $\zeta_i''(t_1) - \zeta_i(t_1)$ and $\zeta_i'(t_1) - \zeta_i(t_1)$. If no such vector can be found, any close neighbor is picked. The trajectories are then traced until their separation reaches $L'(t_2)$, whereupon a new vector is chosen at distance $L(t_2)$, and so on.

In this approach, the highest Lyapunov exponent is defined as

$$\lambda_1 = \frac{1}{t_M - t_0} \sum_{k=0}^{M} \log_2 \frac{L'(t_k)}{L(t_{k-1})} \quad .$$

To estimate the sum of the two largest exponents, one examines pairs of close trajectories as well as the variations of the areas of corresponding triangles (Fig. 4.24b). When choosing new neighboring vectors, one should look for a close parallelism in the trajectories. Here

$$\lambda_1 + \lambda_2 = \frac{1}{t_M - t_0} \sum_{k=1}^{M} \log_2 \frac{A'(t_k)}{A(t_{k-1})} \quad .$$

As a rule, no more than two Lyapunov exponents are estimated by the analogue method.

The usefulness of the above algorithm is illustrated by the following example from hydrodynamics.

One of the systems where a transition from the ordered *laminar* condition to the chaotic turbulent one takes place is Couette–Taylor flow. It is observed between two cylinders of height L and radii a and b (Fig. 4.25). The outer cylinder spins with angular velocity Ω_1 and the inner cylinder is stationary. The fluid motion in this case is governed by a dimensionless parameter, the Reynolds number $\mathrm{Re} = \Omega_1 a(b-a)v^{-1}$, where v is the kinematic viscosity of the fluid.

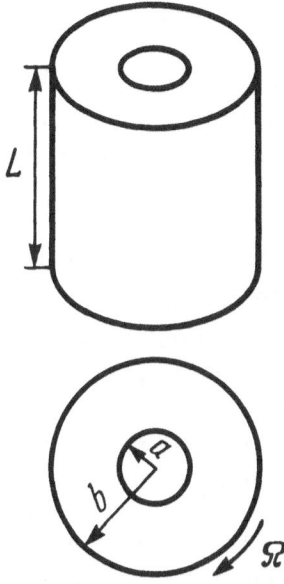

Fig. 4.25. Fluid flow between two spinning cylinders, the favorite object of analysis in synergetics and hydrodynamics

At small Reynolds numbers the fluid motion is laminar and can be given by an explicit formula. Increasing Ω_1 then gives rise to dissipative structures called Taylor vortices. As the angular velocity increases further, the Taylor vortices lose their stability, and the fluid motion becomes periodic (the limit cycle analogue). At still greater Reynolds numbers, a double-frequency condition (the invariant torus analogue) sets in, and finally, at $\mathrm{Re} = \mathrm{Re}_c$, the motion becomes turbulent. Efficient experimental facilities using laser Doppler spectroscopy are available to analyze this flow. By measuring the Doppler shift of a laser beam passing through a fluid, one can estimate one of the velocity components at a given point. Step-by-step measurements of this component yield a series $a_1, \ldots, a_N$.

These data are used to find the correlation exponent and positive Lyapunov exponents at $\mathrm{Re}_c \leq \mathrm{Re} \leq 1.3\,\mathrm{Re}_c$. A typical dependence between the logarithm of $C(\varepsilon)$ (the correlation integral) and $\ln \varepsilon$, with regard to this case, is shown in Fig. 4.26 along with the slope s of this dependence curve. Four regions (typically defined in hydrodynamic and other cases) can be marked fairly clearly on the curve $s = s(\ln \varepsilon)$ as follows [4.28]:

Region A. In this range of resolution scales, cells (that is, cubes with edge ε) are so small that only a few points fall within each of them, which precludes assessment of the probabilities P_i that are used to define dimension.

Region B. In this range, the sampled information is not sufficient to render the attractor's Cantorian structure. Besides, it usually contains a good deal of experimental error, that is, inaccuracy in our estimates of the attractor's points.

Region C. In this range, the points of the curve specify the correlation exponent v_0. To extend region C, we can increase the sample size, refine the accuracy of the experimental data, optimize the selection of variables to be analyzed, and employ special methods to process experimental data.

Region D. The size of refinement cells tends to that of the attractor, and the curve $\ln C = \ln C(\ln \varepsilon)$ is not characteristic of its fractal structure.

The situation is much the same with attractors of higher dimension, but with growing dimension the width of region C diminishes.

These results suggest that in the range of Reynolds numbers covered, v never exceeds 5.4. We can now see that small-mode chaos exists in our infinite-dimensional system. It appears to be a valid assumption that a small-dimensional strange attractor governs this flow. Also, it can be proved that the higher-order Lyapunov exponent is positive.

Experts in the arts and the humanities often mention changing styles in architecture, painting, literature, or ladies' clothing in their writings. Nowadays we commonly assume the changes are due to the *Zeitgeist*, the spirit of the times, or simply to fashion. In the language of mathematics, style can be called a slow variable, and fashion can be seen as the same variable with fluctuations of fairly large amplitude superimposed upon it.

Fashion apparently plays a significant role in the exact sciences too. This will be readily seen if we look through scientific journals:

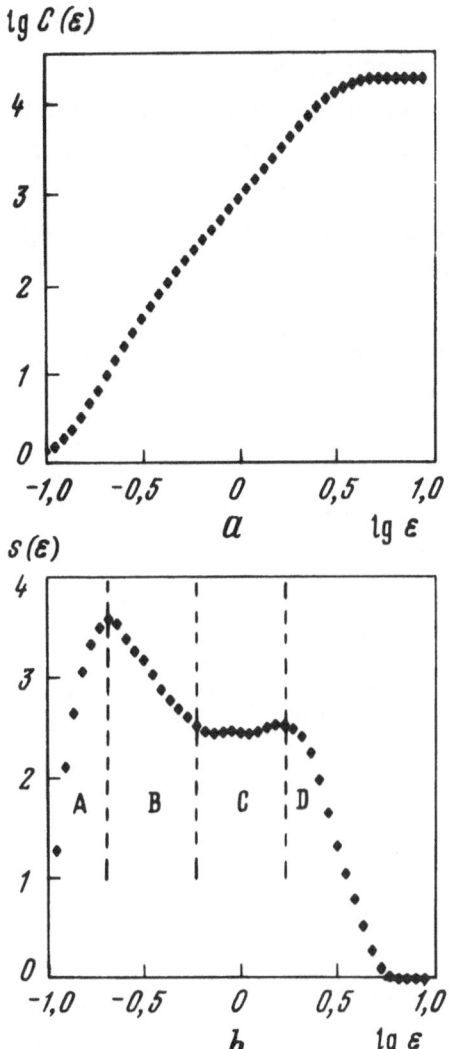

Fig. 4.26. Fractal dimension estimation for Couette–Taylor flow [4.24] based on experimental data ($\mathrm{Re}/\mathrm{Re}_c = 12.4$, where Re_c is the value of the Reynolds number at which the flow loses its stability)

'buzzwords', 'hot' topics, and fashionable examples are easy to spot. About twenty years ago, editors willingly accepted manuscripts containing the word 'soliton'. To publish a paper dealing with dynamic chaos ten years ago, it was sufficient to include an illustration presenting the 'ball of trajectories'. Editors today are eager to publish works concerned with the computation of correlation and Lyapunov exponents. But I don't mean to be sarcastic. The editors' standpoint is understandable, for among the 'informational noise' there may be some momentous and profound contributions to what is believed to be a promising trend. A brilliant study of possibly permanent significance may be hidden under what seems to be an ephemeral work.

Using the algorithms discussed above, one can process any sampling and obtain a collection of numerals. There is an immense and steadily increasing number of problems to which they may be applied. Among the latest ones are computation of correlation exponents for samplings defining the intensity of radiation in the vicinity of a neutron star [4.29], for certain types of solar activity, for daily mean temperature fluctuations and global climatic variations, and for samplings describing business activity; estimation of the Lyapunov exponent for series of infants' cries, and for a sampling defining the arrangement of prime numbers in a series; analysis of irregularities in the Earth's rotation [4.30]; and so on.

From the mere mention of the problems it cannot be readily seen where a knowledge of v and λ_i should lead to important results and where it is merely a tribute to fashion. Very frequently, the studies themselves are not informative in this respect. This is all the more difficult, since only a few numbers can as a rule be derived from studies devoted to the said problems. However, it is not the number in itself, but rather the idea expressed by it and an understanding of how it can be used in future that really count.

One of the architects of chaotic dynamics, David Ruelle, expressed this attitude toward studies lacking the necessary explanatory component in his meaningfully entitled paper 'Deterministic chaos: the science and the fiction'. This is what he said: "Readers of *The Hitch-Hiker's Guide to the Galaxy*, that masterpiece of English literature by D. Adams, know that a huge supercomputer has answered 'the great problem of life, the universe and everything'. The answer obtained after many years of computation is 42. Unfortunately, one does not know to what precise question this is the answer, and what to make of it. I think what happened is this. The supercomputer took a very long time series describing all it knew about 'life, the universe and everything' and proceeded to compute the correlation dimension of the corresponding dynamics using the Grassberger–Procaccio algorithm. This series had a length N somewhat larger than 10^{21}. And you can imagine what happened. After many years of computation the answer came: dimension is approximately $2\log_{10} N \approx 42$" [4.31].

4.6 Dynamical Chaos. Gates of Fairyland

> ... – Those verses of yours, are they good, indeed?
> – Dreadful! – Ivan suddenly said boldly and frankly.
>
> *M. Bulgakov*
>
> *The Master and Margarita*

In what follows, a great deal will be said about the imperfection of existing approaches to the far-from-easy problems involved in analyzing predictions in synergetics. Let us consider first an ideal case. Assume that the process at hand is best given by a one-dimensional mapping:

$$x_{n+1} = g(x_n), \quad n = 1, 2, \ldots$$

For simplicity, let $g(x) = 1 - 2|x|$. Let there be known values $x_1, \ldots, x_M$, where the sample size $M = 10^4$. In this mapping, an arbitrary x_1 taken in the interval $-1 \le x_1 \le 1$ yields a non-periodic sequence, and hence we assume that $x_1, \ldots, x_M$ is a portion of such a sequence. Let us now take another viewpoint. Consider a reverse problem. Given that the values $x_1, \ldots, x_M$ are experimental results, can we use them to reconstruct a dynamical system (a p-dimensional mapping)?

Using the above algorithm, we can see that the attractor's fractal dimension (ν or d_c) is close to unity, while the only positive Lyapunov exponent $\lambda_1 = \ln 2$. Such a sample size allows a sufficiently precise evaluation of both quantities.

We can now presume that under a certain mapping

$$\left\{ x_n^1, \ldots, x_n^p \right\}, \quad n = 1, 2, \ldots,$$

governing the system's dynamics, $p = 1$. That is, there is a unique order parameter, and the process at hand is modelled by the one-dimensional mapping.

For this system, the prediction problem can be stated as follows: using the results of prior observations $x_1, \ldots, x_M$, forecast later values $x_{M+1}, x_{M+2}, \ldots$

To work the problem we can use the following technique. On the basis of the available sample we construct dependence $x_{n+1} = x_{n+1}(x_n)$, that is, the dependence of the $(n+1)$-th element of the series on the n-th one in the plane $\{x_n, x_{n+1}\}$ (Fig. 4.27). We obtain none other than the

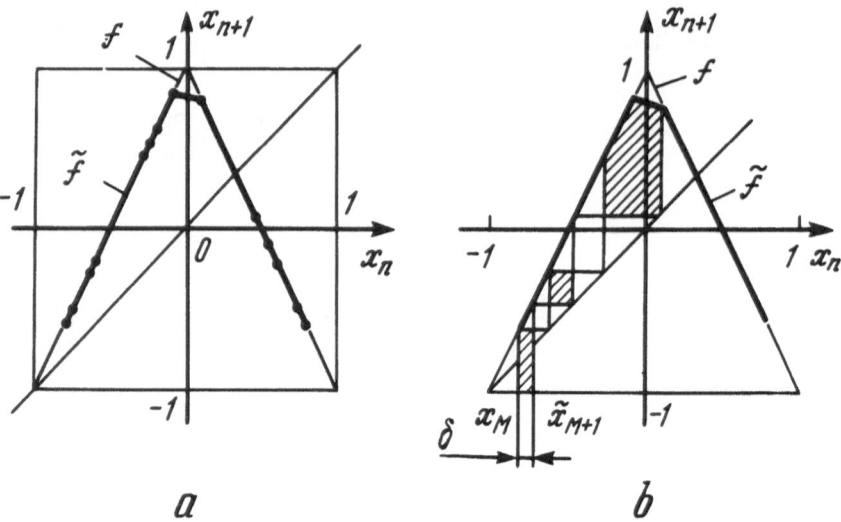

Fig. 4.27. A principal scheme of predictor f construction based on observational data

function f. Naturally, the plane embeds a finite number of points, while for prediction a function is required for every value of x, $-1 \le x \le 1$.

So long as M is large enough, we can employ the standard interpolation algorithm, known since Newton's time, which operates promptly and fairly accurately. Its application results in the function $\tilde{f}$, which enables prediction and is therefore called the *predictor*.

The procedure for making predictions is as follows. Suppose that we know x_M with an accuracy δ. We call this the *measured value*, $\tilde{x}_M = x_M + \delta$. Then the predictand x_{M+1} (Fig. 4.27) is $\tilde{x}_{M+1} = \tilde{f}(\tilde{x}_M)$, predictand x_{M+2} is $\tilde{x}_{M+2} = \tilde{f}(\tilde{x}_{M+1})$, and so on (Fig. 4.27b).

Due to a sensitive dependence on initial conditions, a small prediction error grows stepwise to cause a much greater effect in a finite time. In our case, the error at least doubles, since $\lambda_1 = \ln 2$. Actually, it grows even more, for instead of function g only the approximate dependence $\tilde{f}$ is known. We can therefore make forecasts for k steps ahead, $k < 1 + \log_2 \delta$. (Indeed, assuming that after k steps the error $\delta e^{\lambda_1 k}$ is compatible with the attractor's typical size S we get $\delta e^{\lambda_1 k} = S$, $k = (\ln S - \ln \delta)/\lambda$. In our case, $S = 2$, $\lambda_1 = \ln 2$, $k = 1 - \log_2 \delta$.)

Proceeding in a similar way, we can produce predictands for a vector **x**. This was done by the American scientist Farmer for several differential equations and experimental data describing one of the velocity components for a turbulent fluid flow [4.32].

The Takens theorem involves using an infinitely long series of ideally accurate data sampled at intervals Δt. Suppose that we can manipulate the sampling time Δt any way we like. Which is the best way to do it? The natural requirement for a sample of a given length N is that it provide a highly accurate description of an attractor.

Several approaches for this application have gained recognition. One of them is quite simple. If Δt is small enough, the projection of an attractor onto some plane in a ζ-space is close to a diagonal (Fig. 4.28a). If Δt is large, the attractor may become too complex. In an optimal case, the attractor should be both simple enough and set up to best advantage (Fig. 4.28b). To acquire the most complete information, the sequential ζ-vectors must be 'supremely independent'. This is accomplished with the help of the autocovariation function

$$b(\tau) = \frac{1}{T}\int\limits_0^T x(t)x(t+\tau)\,dt\,,$$

where Δt is chosen to correspond to the first zero $b(\tau)$. We can also introduce a more complicated function called the *mutual information*. To compute Lyapunov exponents, we pick Δt using a different procedure. As we see, Δt can be disposed of quite reasonably [4.33, 34].

It is not so easy with the sample size N. A computation of the correlation integral was suggested by Grassberger and Procaccio [4.26] in the same work in which they gave test results for one- and two-dimensional mappings, systems of differential equations, and delay

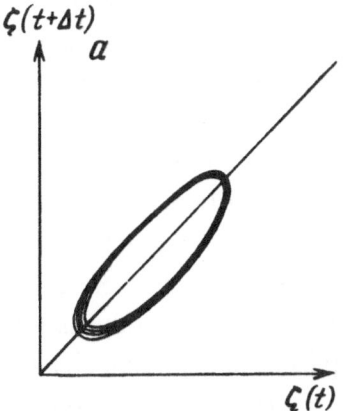

Fig. 4.28a. A typical behavior pattern for the case when Δt is too small

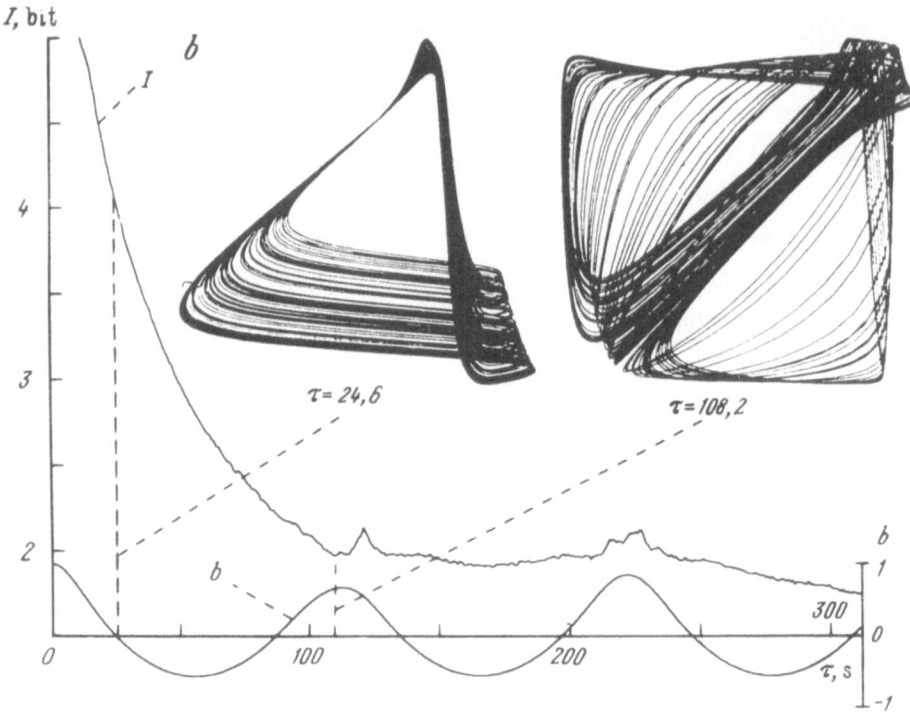

Fig. 4.28b. A reconstruction of an attractor using experimental data for the Belousov–Zhabotinskii oscillatory chemical reaction

equations. The latter class of mathematical models is now widely used in ecology, nonlinear optics, and immunity theory. They take the form

$$\dot{u}(t) = f\big(u(t - \tau)\big) - g\big(u(t)\big).$$

There exist possibly an infinite number of degrees of freedom for this equation; to provide the solution, a function must be defined for the whole interval from 0 to τ. At large delay times τ, such equations describe chaotic processes. The fractal dimension of the corresponding attractor usually increases with τ.

Grassberger and Procaccio used relatively small samples comprising 10 000–30 000 elements to estimate exponents with fairly good accuracy. The results generated much optimism. Indeed there is a great host of interesting problems (like those concerning the solar and terrestrial dynamos, and climate and weather attractors) where the hypothesis that deterministic chaos exists is established. Samples of rather small size are used in each of these cases. Thus, to analyze climatic variations during the last thousand years they used samples not exceeding 500–700 elements. To tackle the geodynamic attractor problem (the non-uniformity of the Earth's rotation during recent centuries) they employed 700 elements. In the case of attractors for weather systems, no more than 60 000 daily observation results were available. There is presently a lively debate in scientific journals like *Nature* and *Journal of the Atmospheric Sciences* over whether or not such samples are sufficient to draw substantiated conclusions concerning these phenomena.

For example, the following reasoning is advanced. Consider a v-dimensional attractor within a p-dimensional space. To define its geometry and evaluate the probabilities P_i involved in the formula for generalized dimensions D_q, at least n points must fall within each p-dimensional cube containing the attractor's points. In the course of reconstructing the attractor, each of the components of the v-vector assumes all values from ξ_{max} to ξ_{min}. Let $\xi_{max} - \xi_{min} = L$. A p-dimensional cube with edge L embeds $(L/\varepsilon)^p$ cubes with edge ε, and consequently the sample size must be $N \cdot (L/\varepsilon)^p$, that is, $p \approx \ln N$. The largest dimension of the attractor that can be estimated using a sample of N elements is logarithmically dependent on N. This usually sounds convincing.

A central question now is: how can we find the coefficient governing this dependence? The authors of one well-known study claim that to estimate the Hausdorff dimension d_H to within a few percent accuracy, one needs a sample larger than 42^α, where α is the integral part of d_H or the integral part plus one [4.35]. Thus, to analyze an attractor with fractal dimension 3 one needs over 100 000 elements, and fractal dimension 6 correspondingly requires over 10 billion elements. If this is really so, then why does the use of far smaller samples in some cases yield good results? No clear and simple answer to this question has been provided so far.

Another problem has to do with the effects of noise and finite precision. The definition of fractal dimensions implies that results are derived in the limit as $\varepsilon \to 0$ or $N \to \infty$ [4.36]. It is naturally prohibitively difficult to evaluate them directly through the analysis of particular data. Let us now see how things develop. In an ideal case, reducing the scale of a strange attractor without bound we observe one and the same self-reproducing structure. However, if a system is noisy, only several scale levels of such a structure are observable. If the noise is too great or the measurement accuracy too low, the reconstructed attractor may completely lose its unusual geometric features. Therefore we are forced to evaluate the noise level for each individual system and to devise a procedure for estimating the dimension which is based on the final sample.

For many experimentors, the favorite methods of data processing are smoothing and numerical differentiation, both of which are risky in terms of dynamical system theory. Depending on the parameters of the algorithms at hand, they may either reduce or exaggerate the reconstructed attractor dimension.

In each individual case the solution of these problems calls for a detailed analysis of all computational procedures employed as well as investigation of many methodological issues. Based on my own experience in handling observational data related to the geodynamic attractor problem, weather variations, and numerical experimental modeling of three-dimensional turbulent fluid flows, I would go so far as to say that in each case even the statement of the problem of how to give a numerical description of the attractor should be individual [4.30, 37].

In the absence of a meticulous treatment of each particular system from the standpoint of the underlying attractor, all we can expect is a rush of remarkable 'findings' followed by corresponding disavowals. As an example, let us consider the case of the weather attractor. The problem to be solved is whether or not variations of the average daily temperature are described by a dynamical system containing a strange attractor. The course of the discussion of this problem in the literature is very interesting. Initially, the hypothesis was advanced that $v \approx 3 + 4$ [4.38] for a reconstructed attractor, based on an analysis of samples containing several thousand elements. This was disputed later by those who claimed that $v = 6 + 8$, based on the analysis of neighboring samples [4.39]. Finally, there appeared papers where the weather attractor's correlation exponent was assessed as $v = 10 + 22$.

Things developed similarly when the problem of the climatic attractor was discussed.

To be able to judge the applicability of these data to the problem of finding the desired information or to assess the accuracy of the methods employed, one must have access to the samples in question (commonly, Berlin observatory databases). The samples were inaccessible to me and my co-worker A. B. Potapov from the Keldysh Institute of Applied Mathematics. However, we made the necessary computation using observational data donated by the Clementine observatory in Prague (with a sample size $N \approx 6 \cdot 10^4$). Ignoring reconstruction and with only preliminary sample processing, the value of v obtained was computed to be close to 10, though the accuracy was not very high. A short flat portion of the curve $s(\varepsilon) = (d \ln C)/(d \ln \varepsilon)$ actually governs v.

Now we try another approach. In computing the correlation integral, let us see what neighbors the randomly chosen point x_i has (Fig. 4.29). Where does the largest contribution to $C(\varepsilon)$ come from? The answer is simple – the largest contributors are the trajectory neighbors. Clearly they have nothing to do with the attractor's structure. If Δt is very small, and the number of trajectory neighbors is very large compared to the remaining points, then the attractor is observed to resemble a one-dimensional curve ($v = 1$). To examine its structure, we first have to remove the nearest neighbors of the point on its trajectory. The procedure for doing so is well known to experts [4.40]. Once it is performed, the flat portion of the curve $s(\varepsilon)$ vanishes. It becomes clear that the limiting value of v for this sample was due merely to an imperfection of the procedure.

By analyzing a far larger sample (so far lacking in national meteorological centers), it may be possible to obtain the true value of v. Using a still larger database, investigators could have estimated Lyapunov exponents and constructed predictors. Well, maybe not – there is much to be said on both sides.

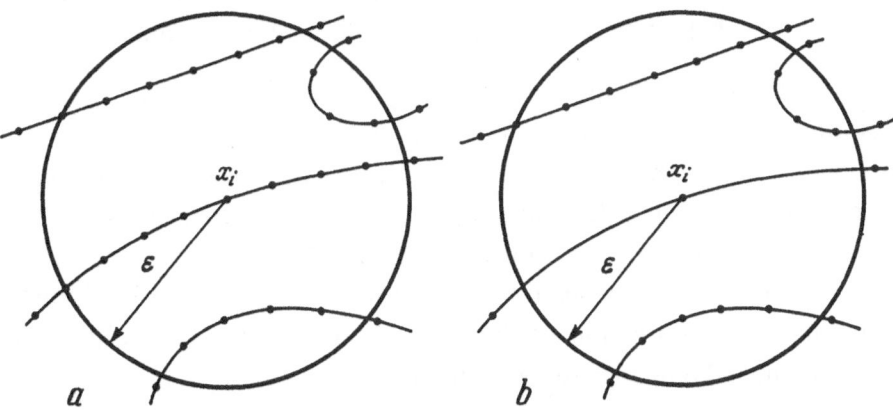

Fig. 4.29a,b. Estimation of the correlation exponent of the weather and climate attractor: **(a)** points falling within the sphere of radius ε, **(b)** the situation after removal of trajectory neighbors of point x_i

Before leaving the discussion, we mention one more point concerning the problems of weather and climate attractors and some other problems of this type. In these cases we are dealing with multi-frequency oscillations involving a continuous smooth spectrum, where it is impossible to isolate the 'most important' frequencies (except for the annual cycle, of course). Roughly speaking, we cannot say where the line distinguishing weather and climate is to be drawn. This point represents a great complication for the analysis of the data.

After what has been said in this section, the problem of forecasting weather, as well as predicting the dynamics of other open nonlinear systems, appears to be a tangle of insoluble riddles.

Some researchers might think that research in this field has come to a dead end, and that those who performed it would do better to get another job. Such a view is sometimes expressed at scientific conferences.

If the use of a synergetic approach to the solution of forecasting problems does not seem attractive, or if the reader is not impressed by the fact that some of these approaches have already met with reasonable success in overcoming concrete problems, then there is only one remaining alternative, namely, to relate what we specialists in synergetics are looking forward to.

At any rate, the future of science – and not just of science – is based not only on what has been done in the past and what goes on at present but also on our ideas about the future.

The simplest way to put this is through formulation of different (sometimes mutually exclusive) viewpoints on the causes of current difficulties. These may be deliberately simplified or exaggerated.

1. Forecasting temperature at one point is an unpromising under-taking. It is not daily mean temperature that you forecast, but some functionals, e.g., pressure and temperature averaged over some region, or functionals related to time-averaging. Perhaps the idea of forecasting should be reduced to the following scheme: "if January is snowy and February cold, July will be dry", which naturally implies a preliminary quantitative specification of the notions 'dry', 'snowy', 'cold', error evaluation, and other necessary operations.

2. The Lyapunov exponent defines the rate of divergence of close trajectories averaged over a large time interval. However, it does not give the behavior of individual trajectories in separate regions of phase space.

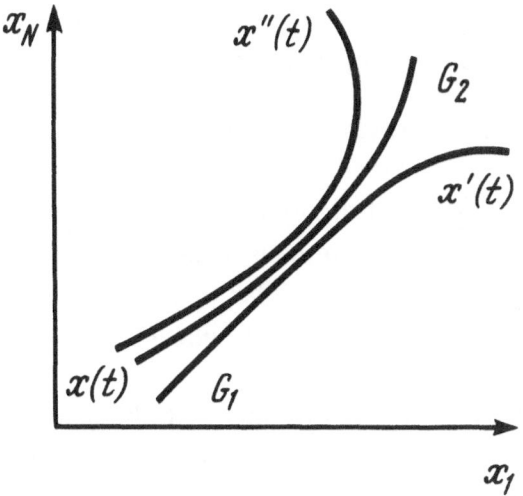

Fig. 4.30. Behavior of close trajectories in different regions of a phase space

Why not make use of the situation images in Fig. 4.30 from time to time? In region G_1 close trajectories converge slowly, while in region G_2 they diverge rapidly.

If our system falls in region G_1, we have everything we need to make forecasts until after it leaves this region. Forecasting in this case can be done in the following manner: if pressure over region A falls within range I_1, temperature within range I_2, humidity within range I_3, and so on, then in time τ things will develop as follows ... As a matter of fact, some phenomena studied in physics can be predicted in this way.

3. An elegant technique whereby noisy cycles can be treated has alreayd been discussed.

4. New methods of accounting for a priori information are called for. Indeed, a 'point-blank' computation of fractal dimension and Lyapunov exponents is often inefficient. At the same time, correction of a trend, smoothing, and – conversely – complication of different spectrum regions considerably alters the structure of reconstructed attractors.

It is conceivable that the modified attractor carries along the most important information about the system. Some time ago a similar program dealing with the isolation of slow variables was popular in meteorology. A large effort was made to validate the hypothesis that the state of both the atmosphere and the ocean at a small number of key points enables a fairly precise characterization of atmospheric conditions as well as their forecast. But the work in this direction yielded no appreciable results. Success may be achieved later.

Two other approaches that might provide a clue to the problem are now under intensive development in synergetics.

The first one arose from the theory of algorithms. Assume that we have at our disposal a giant computer with enormous memory and speed. Sure enough, it can be used to simulate processes taking place in many nonlinear dissipative systems existing in nature whose behavior we desire to predict.

Ultimately, what is a computer? It is a machine that changes the contents of memory locations at instants Δt, $2\Delta t$,... in accordance with preset rules. Each of the memory locations may be in any one of a finite number of states. The rules governing how the contents change are themselves contained in memory locations.

There are usually several algorithms whereby one or another operation is performed. For example, assume that an addition algorithm for integers written in the form of binary numbers (i.e., numbers x written as strings $a_n a_{n-1} \ldots a_1 a_0$, where

$$x = a_0 + 2 \cdot a_1 + 2^2 \cdot a_2 + \ldots + 2^{n-1} \cdot a_{n-1} + 2^n \cdot a_n$$

and $a_i \in \{0,1\}$) is programmed into the computer. Suppose we want to multiply x by 2^k. The simplest, though time-consuming, way to do it is by adding x to itself 2^k times. Time can be saved by shifting the set $a_n \ldots a_1 a_0$ to the left by k elements. Quantity $2^k x$ is the number corresponding to the string

$$a_n \ldots a_1 a_0 \underline{00 \ldots 0} \ .$$
$$k \text{ zeros}$$

Thus 2^k operations are reduced to one. Algorithms that can be executed using a shorter detour are called *reducible*, in contrast to *nonreducible* ones for which no such detour can be found. In other words, reducibility of an algorithm is closely connected with predictability. If an initial algorithm defines a system's state after N_1 steps, and a second one drives the system into this state after N_2 steps, $N_2 < N_1$, the latter can be looked upon as a prediction algorithm for a system whose evolution is imitated by the first algorithm. This connection attracted the attention of the American researcher Stephen Wolfram. The hypothesis he advanced is as follows. Problems where application of the methods of theoretical physics have met with prominent success are classified as computationally reducible ones. It is typical of such problems either that several degrees of freedom can be discerned which produce a system's dynamics or that many identical subsystems can be found. As a result, the approaches to their solution which are usually efficient are averaging and conversion to comparatively simple relationships. However, turbulent fluid flows, biological evolution, and many aerophysical phenomena are classified as computationally nonreducible systems. The only way to analyze their evolution is by simulation, that is, by direct computation of systems that are comparable in complexity. In this context, it is probably no use counting on simplified efficient models. The attractor conception also appears impractical. From different initial data one can reach radically different steady states. This paradoxical development is illustrated by a game called 'Life' devised by the British mathematician Conway [4.42]. Playing Life enables one to see how simple the rules governing computationally nonreducible systems can be.

Life is played on an infinite flat array of square cells. Discrete time ($t = 1, 2, ...$) is used. A cell may be either living or dead. The state of a cell at instant $t + 1$ is determined by the states of its neighbors at instant t (each cell has eight neighbors; it shares edges with four, and vertices with the other four). The rules are very simple. If a cell is dead at instant t, it becomes alive at instant $t + 1$ provided that just three of its eight neighbors were alive at moment t. A living cell becomes dead if it has less than two or more than three neighbors. That's all. One can trace the evolution of the simplest configurations with just a pencil and a sheet of checked paper. A computer enables one to track the behavior of larger configurations over longer times.

There is a very close similarity between this discrete system and the processes taking place in nonlinear dissipative media. However, at least three important questions naturally arise. What basic types of structures (i.e., patterns in the behavior of groups of squares in the field of view over large times) can exist within a given discrete system? What laws govern the organization of the structures? Can structures interact and, if so, what comes out of this interaction?

The simplest structures are stationary, i.e., time-independent ones. Some of them are displayed in Fig. 4.31a. The basic idea is that these structures are localized. If they are separated from each other by two or

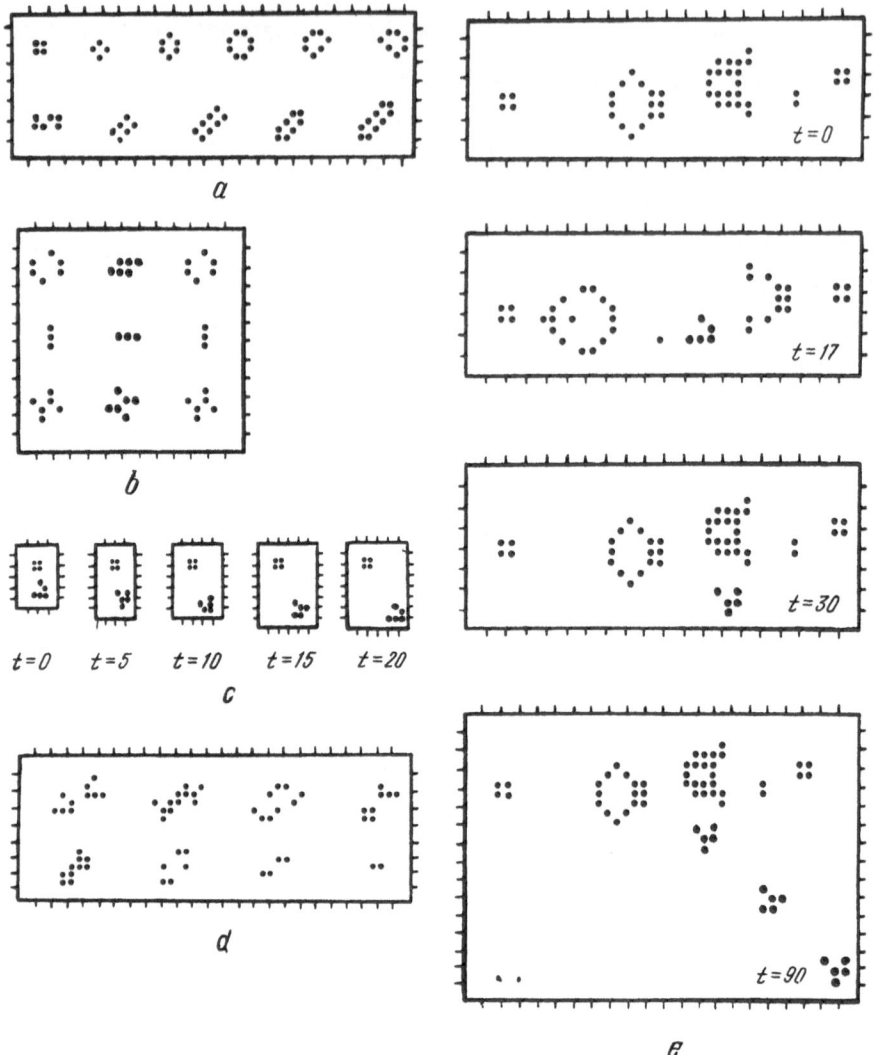

Fig. 4.31a–e. Various configurations occurring in the game of Life: (**a**) stationary structures, (**b**) examples of cycles, (**c**) the glider, (**d**) collision between two gliders, (**e**) the catapult or the glider gun (living cells are marked with dots and the rectangles are marked at two-square intervals)

more dead cells then they exert no influence on one another. It is safe to say that stationary structures reproduce themselves with each step in time. But what will happen to them after 2, 3, or p steps? For simplicity, we shall call such configurations cycles. They describe periodic processes in groups of cells.

Binary cycles are shown in Fig. 4.31b. This figure should be examined line by line. Since binary cycles are localized, they all fall in the same region. The evolution of various groups often involves the binary cycle on the second line, called 'the semaphore'.

Let us see what other types of order the game of Life displays. Figure 4.31c shows a remarkable configuration called 'the glider' (the stationary structure above it is shown as a reference mark). The glider reproduces itself after every four steps and also shifts down and to the right by one square (clearly there can be gliders moving along either diagonal in either direction). The figure shows four different glider phases.

A collision between two gliders or between a glider and a stationary structure results in their annihilation (Fig. 4.31d). Other collisions may produce a collection of stationaries and semaphores. The larger the area occupied by a group, the more complex its behavior. This is why configurations that grow in space without bound are interesting. One of them, called 'the catapult' or the 'glider gun', was announced in 1970 by Gosper. Figure 4.31e shows that after every 30 steps the catapult reproduces itself and emits a glider. The gun shoots forth a stream of gliders.

A configuration like the glider gun is of fundamental importance. Indeed, a computer can be assumed to be a finite set of simple logic gates carrying out AND, OR, and NOT operations and wired in some specific way to enable the transfer of pulses coding 0–1 sequences.

The generator of such pulses in the game of Life is the glider gun. Its presence in the flow can naturally be interpreted as 1 and its absence as 0. The collision of gliders resulting in annihilation produces the NOT state with two flows set at right angles (because a glider traveling with one flow is found in a given location, upon collision another glider traveling with another flow vanishes in this location). The construction of other elements in this game is more complicated. However, it is possible, and Life turns out to be equivalent to a *universal computer*. Under certain specific conditions, processes of arbitrary complexity can evolve, and can be simulated again on a computer.

An alternative approach to forecasting, now finding a growing number of adherents, involves neural networks. A striking difference between the way the human brain works and the operation of modern computers is the following. We rely on our innumerable informal intuitive notions when a quick assessment of a situation is needed or an effective decision must be made. Furthermore, we have countless opportunities to learn by picking what we need to solve our problems out of the flow of information around us. The latter ability has to do with the great gift of forgetting, and preserving in memory only key fragments and important details.

This is why many aspects of mental work within a child's reach are immensely difficult for a computer. Among them are recognition of faces, learning to speak, solution of simple non-formalized problems, generalization, and so on.

Nevertheless, the last few years have seen the emergence of a new class of computing systems, called *neurocomputers,* designed to tackle problems beyond the capabilities of conventional computers [4.43]. Roughly speaking, neurocomputers are structures that are capable of changing their own structure in conformity with incoming information from the tasks at hand. Actually, these machines reconstruct themselves anew to reflect the reality bounded by their perception. This opens up new possibilities for training: development, searching for analogies, reconstruction of entities from separate fragments, and scientific prediction. Indeed, suppose that there is a complex system for which neither the order parameters nor the defining relationships are known, but there is an opportunity to track its behavior over a fairly long period of time. It can be hoped that a neurocomputer will succeed in picking relevant information and finding interrelations that can be readily 'sieved through' the algorithms discussed above (and many others), enabling it to construct an adequate computing structure. Such an approach is now an area of active research in connection with earthquake forecasting, data processing, and the construction of intelligent robotic systems capable of analyzing the course of certain processes.

We can anticipate that there will be considerable progress in this area in the near future. By that time, probably, it will become clear to everybody that specialists in predicting complex system behavior need not abandon their pursuit in favor of more practical jobs, that the ideas of synergetics are fruitful, and that all the great efforts made to carry on these studies in the 1990s were not in vain.

Acknowledgements

It is my pleasant duty to thank A. A. Samarskii, S. P. Kurdyumov, A. B. Potapov, T. S. Akhromeyeva, and S. V. Ershov for their support and assistance.

References

4.1 M. Kline: Mathematics and the search for knowledge. New York: Oxford University Press, 1985

4.2 R.P. Feynman, R.B. Leighton, M. Sands: The Feynman lectures on physics. Reading, MA: Addison–Wesley. Vol. I, 1963; Vol. II, 1964; Vol. II, 1965

4.3 V.I. Oseledets: The multiplicative ergodic theory. Characteristic Lyapunov exponents for dynamical systems. Trans. Moscow Math. Soc. 19, 179–210 (1968) (in Russian)

4.4 G. Benettin, L. Galgani, A. Giorgilli, J.M. Strelcin: Lyapunov characteristic exponents for smooth dynamical systems and for Hamiltonian systems: a method for computing all of them. Parts 1, 2. Meccanica 15(1), 9–20, 21–30 (1980)

4.5 E.P. Wigner: Symmetries and reflections. Bloomington: Indiana University Press, 1967

4.6 H. Haken: Synergetics. Berlin: Springer-Verlag, 1978

4.7 G. Nicolis, I. Prigogine: Self-organization in nonequilibrium systems. From dissipative structures to order through fluctuations. New York: Wiley, 1978

4.8 I. Prigogine, I. Stengers: Order out of chaos. Man's new dialogue with nature. London: Heinemann, 1984

4.9 G. Iooss, D.D. Joseph: Elementary stability and bifurcation theory. New York: Springer-Verlag, 1980

4.10 T.S. Akhromeeva, S.P. Kurdyumov, G.G. Malinetskii, A.A. Samarskii: Nonstationary dissipative structures and diffusion-induced chaos in nonlinear media. Physics Reports 176(5, 6), 189–372 (1989)

4.11 R. Gilmore: Catastrophe theory for scientists and engineers. New York: Wiley, 1981

4.12 N.N. Moiseev: Alternatives of development. Moscow: Nauka, 1987 (in Russian)

4.13 C. Foias, G.R. Sell, R. Temam: Inertial manifolds for nonlinear evolutionary equations. J. Different. Equations 73(2), 309–353 (1988)

4.14 E.N. Lorentz: Deterministic nonperiodic flow. J. Atm. Sci. 20, 130–141 (1983)

4.15 J.D. Farmer, E. Ott, J.A. Yorke: The dimension of chaotic attractors. Physica D, 7(1–3), 153–180 (1983)

4.16 J. Kaplan, J. Yorke. In: Functional differential equations and the approximation of fixed points. Proceedings, Bonn, July 1978. Lecture Notes in Math. 730, p. 228. Berlin: Springer-Verlag, 1978

4.17 G.G. Malinetskii, A.A. Ruzmaikin, A.A. Samarskii: Model of solar activity long-range variations. Preprint of Keldysh Inst. Appl. Math., Russian Acad. Sci., 170. Moscow, 1980 (in Russian)

4.18 A.E. Cook, P.H. Roberts: The Rikitaki two-disk dynamo system. Proc. Cambridge Phil. Soc. 68, 547–569 (1970)

4.19 S.V. Ershov, G.G. Malinetskii, A.A. Ruzmaikin: A generalized two-disk dynamo model. Geophys. Astrophys. Fluid Dynamics 47, 251–277 (1989)

4.20 P. Collet, J.P. Eckmann: Iterated maps on the interval as dynamical systems. Basel: Birkhäuser, 1980

4.21 M.V. Jakobson: Absolutely continuous measures for one parameter families of a one-dimensional map. Commun. Math. Phys. 81(1), 39–88 (1981)

4.22 J.D. Farmer: Sensitive dependence on parameters in nonlinear dynamics. Phys. Rev. Letters 55(4), 351–354 (1985)

4.23 S.V. Ershov, G.G. Malinetskii: On the solution of the inverse problem for the Perren–Frobenius equation. J. Comput. Math. and Math. Phys. (in Russian) 28(10), 1491–1497 (1988)

4.24 F. Takens: Detecting strange attractors in turbulence. In: Dynamical systems and turbulence. Proceedings, Warwick, 1980. Lecture Notes in Math. 898, pp. 366–381. Berlin: Springer-Verlag, 1981

4.25 H.G.E. Hentschel, I. Procaccio: The infinite number of generalized dimensions of fractals and strange attractors. Physica D, 8(3), 435–444 (1983)

4.26 P. Grassberger, I. Procaccio: Measuring the strangeness of strange attractors. Physica D, 9(1, 2), 189–208 (1983)

4.27 A. Wolf, J.B. Swift, H.L. Swinney, J.A. Vastano: Determining Lyapunov exponents from a time series. Physica D, 16(3), 285–317 (1985)

4.28 A. Brandstater, H.L. Swinney: Strange attractors in weakly turbulent Couette–Taylor flow. Phys. Rev. A, 35(5), 2207–2220 (1987)

4.29 H. Atmanspacher, H. Scheingraber, W. Voges: Global scaling properties of a chaotic attractor reconstructed from experimental data. Phys. Rev. A, 37(4), 1314–1322 (1988)

4.30 G.G. Malinetskii, A.B. Potapov, S.H. Gizzatulina, A.A. Ruzmaikin, V.D. Rukavishnikov: Dimension of a geomagnetic attractor from data on length of daily variations. Phys. Earth and Planetary Interiors 59, 170–181 (1990)

4.31 D. Ruelle: Deterministic chaos: the science and the fiction. The Claude Bernard Lecture 1989. Proc. Roy. Soc. London A427, 241–248 (1990)

4.32 J.D. Farmer, J.J. Sidorowich: Predicting chaotic time series. Phys. Rev. Letters 59(8), 845–848 (1987)

4.33 A.M. Fraser, H.L. Swinney: Independent coordinates for strange attractors from mutual information. Phys. Rev. A, 33(2), 1134–1140 (1986)

4.34 T.S. Akhromeeva, S.P. Kurdyumov, G.G. Malinetskii, A.A. Samarskii: Nonstationary structures and diffusion-induced chaos. Moscow: Nauka, 1992 (in Russian)

4.35 L.A. Smith: Intrinsic limits on dimension calculation. Phys. Letters A, 133(6), 283–288 (1988)

4.36 Yu.A. Kravtsov: Randomness, determinateness, predictability. Sov. Phys.-Uspeckhi 32(5), 434–449 (1989)

4.37 G.G. Malinetskii, A.B. Potapov, V.G. Primyak: On the possibility of description of viscous fluid turbulent flows by a finite-dimensional attractor. Proc. AN SSSR-Doklady 316(5), 1101–1106 (1991) (in Russian)

4.38 K. Fraedrich: Estimating the dimensions of weather and climate attractors. J. Atm. Sci. 43(5), 419–432 (1986)

4.39 K. Fraedrich: Estimating weather and climate predictability on attractors. J. Atm. Sci. 44(4), 722–728 (1987)

4.40 J. Theiler: Spurious dimension from correlation algorithms applied to limited time-series data. Phys. Rev. A, 34(3), 2427–2432 (1986)

4.41 S. Wolfram: Universality and complexity in cellular automata. Physica D, 10(1–2), 1–35, 1984

4.42 E.R. Berlekamp, J.H. Conway, R.K. Guy: Winning ways for your mathematical plays. London: Academic Press, 1982

4.43 E. Domany. Neural networks: a biased overview. J. Stat. Phys. 51(5/6), 743–775 (1988)

5. The Information-Theoretic Approach to Assessing Reliability of Forecasts

V. A. Lisichkin

5.1 Assessing Forecasts

For the past 20–25 years, forecasting has ceased to be the preserve of science and has become an area of professional interest to those engaged in industry, engineering, technology, and other fields. To be certain that an activity is effective (that is, useful or beneficial to society) one needs a means to assess the validity of its results. What can forecasting provide? Usually it gives us more or less adequate information about the future or a set of signals containing such information [5.1].

In the international literature on the subject, two approaches to interpreting this information can be found, the semantic and the statistical. The semantic approach has been treated at great length by Zeman [5.2], Morozov [5.3], Ashby [5.4], Kharkevich [5.5, 5.6], Miles [5.7], Ackoff and Emery [5.8], and so on. The statistical approach is based on fundamental studies by Shannon [5.9], Reed and Gubbins [5.10], Canyon and Steimon [5.11], Bayes, and others, and discussed in numerous monographs and articles by modern authors.

We shall not touch upon the virtues and disadvantages of these approaches as applied to the problem at hand. Let us just say that we have tried to devise an information-theoretic method of investigation with the intention of giving a better interpretation of prediction results. Using this approach, we can restate the problem as that of assessing the validity of a signal transmitted via a noisy communication channel.

5.2 Forecasting as the Subject Matter of Information Theory

Forecasting involves processing information acquired from certain sources and systematized in a certain way. Therefore it can be looked upon as data transmission via communication channels. As regards technical communication channels, they fit the well-known Shannon theorems [5.9], for example, the following theorem:

Given a data source x whose entropy per unit time is $\tilde{H}(x)$, and a channel with capacity C, and provided $\tilde{H}(x) > C$, no signal can be transmitted undelayed and undistorted, in any encoding mode. Conversely, for $\tilde{H}(x) < C$, any message of sufficient length can be encoded and transmitted undistorted and noise-free with probability close to 1.

When interpreting prediction as signal transmission it is natural to generalize the notion of a communication channel. Suppose that a channel links two points (r_1, t_1) and (r_2, t_2) of a four-dimensional continuum. We consider a specific case where coordinates x, y, z are fixed values and time is a variable. Communication between (r_1, t_1) and (r_2, t_2) can be assumed to be data transmission either from the past to the present $(t_1 < t_2)$ or from the future to the present $(t_1 > t_2)$. Of course, the channel can only be physically realized with data transmitted from the past to the future. However, we shall examine the case of transmission from the future on the assumption that it is identical to forecasting.

To specify the state of a system at point (r_1, t_1), where t_1 is some instant in the future, let us introduce the notion of a 'predictability demon' operating as a data transmitter from the future to the present. On a superficial view, the idea of the predictability demon is similar to that of the demon proposed by Laplace. The Laplace demon has absolute knowledge about the initial conditions in both future and present time instances, and also knows the laws according to which development may proceed, whatever they might be, and therefore it can make an absolutely precise prediction for an arbitrarily long time ahead. That the existence of the Laplace demon is impossible is more or less clear. This is proved, for example, with the help of the negentropy information-theory principle. In contrast to the Laplace demon, the predictability demon controls information about a system's future state and conveys it to us in the present.

Clearly, a predictability demon that could define future states with absolute precision, avoiding distortion, would be nothing but a converted Laplace demon, and hence cannot exist. However, we assume that it defines a future state with some uncertainty, so that the information it originally possesses is not infinite, that is, $\Delta I < \infty$, and that some amount of information is lost in transmission due to noise, and therefore we can regard the predictability demon as a suitable prediction model.

It is practically impossible to give a perfect representation of a system's future state by forecasting. Information coming from the future is partially lost due to what we shall call prediction noise.

Consider a system composed of an information source X, a communication channel K, and a receiver Y. The source of information constitutes a system whose state in the general case is described by several random variables $x_1, x_2, \ldots, x_s$ with probability densities $f_x(x_1, x_2, \ldots, x_s)$. We shall restrict our discussion to a simple case of a random quantity X having probability density $f_x(x)$ and a receiver characterized by random Y with probability density $f_x(y)$.

Suppose that the communication channel K lies within a time continuum, rather than a space one. That is, the information source X whose state is defined with some probability by a demon is separated from the receiver by time length t. The state of the receiver Y is a prediction of the state of the source X. Thus, the prediction time length corresponds to that of the communication channel. Several ratios relating prediction parameters to each other can be obtained.

Provided a channel is noise-free, the amount of information input to system Y is the entropy $H(x)$ of system X. In fact, however, it is smaller by $H(x/y)$, the conditional entropy of X.

Expressing $H(x)$ in terms of a one-dimensional probability density $f_x(x)$ and $H(y/x)$ as a two-dimensional density $f_{xy}(x,y)$ and conditional probability $f(y/x)$, the information flow from X to Y takes the form

$$I_{X \to Y} = -\int_{-\infty}^{+\infty} f_x(x) \log f_x(x) \, dx + \int \int_{-\infty}^{+\infty} f_{xy}(x,y) \log f(x/y) \, dx \, dy.$$

It is extremely difficult to estimate a system's probability density $f(x)$ as well as the conditional density $f(x/y)$ for any future time. However, using the above expression for $I_{X \to Y}$ it is possible, for example, to compute the largest possible amount of information, max $I_{X \to Y}$, transmitted via a noisy communication channel. This quantity represents the maximum information one can get about the future at instant t under given conditions and using a specific forecasting method. The nature of the method actually determines the quality of the channel.

In the general case, acquisition of the maximum information is a far-from-easy procedure. It depends on the origin of the noise, on the probability of distortion of symbols in a noisy channel, and so on. It is also relevant whether the noise sources are interrelated, and what are their conditional probabilities and correlation functions. The problem is solvable, however, under the assumption that the noise has a white frequency spectrum and is normally distributed, and that the probability of distortions μ is a function of channel quality and length, that is, ultimately a function of the nature of the prediction model and the prediction lead-time β. Probability μ relates β and time t: $\mu = \mu(t, \beta)$. The largest prediction lead-time, or alternatively the predictability horizon, is determined the nature and quality of the prediction model. This can be assessed by setting $\max I_{X \to Y} = 0$, that is, by taking a limiting case where information is completely lost in the communication channel, making estimation of the future state of the system impossible. Let us try to apply this approach to one of the very simplest cases.

5.3 An Example

Let us try to make a prediction regarding two contradictory events that are members of finite sets of events that may occur by some definite time. For example, the contradictory alternatives may be the discovery and non-discovery of the nature of gravitation by the year 2000[1]. We designate the 'non-discovery' as 0 and the 'discovery' as 1. Let a demon assess the probabilities of these two events as P and $1 - P$, respectively, and convey to us the derived information encoded in the binary symbols 0 and 1. In the course of the transmission, each symbol can be distorted, i.e., substituted (0 for 1 or 1 for 0) with probability $\mu(\beta, t)$. As a result, the 'non-discovery' is predicted with probability r and the 'discovery' with probability $1 - r \neq 1 - P$.

---

[1] The legitimacy of an example about *a discovery to be made by 2000* may seem dubious. Firstly, a case dealing with a *discovery* is hard to describe in terms of the statistical approach, since its probability *P is not a measurable quantity*. It would be easier to suggest a more straightforward example, say, where P is the 'real' proportion of bespectacled or AIDS-infected people and r the predicted share. However, the proposed case is instructive in that it allows application of a subjective probability assessment theory, and thereby extension of the approach examined here. Secondly, a connection can be traced between the approach here and the Bayesian approach. An elegant prediction model can be constructed using Bayesian relationships. But this is something for future research work.

To summarize, a forecast can be looked upon as a transmission of elementary symbols 0 and 1 from source X to receiver Y via a noisy communication channel K. We can compute data loss due to transmission, or conditional entropy $H(y/x)$. It depends merely on the probability of distortion $\mu(\beta,t)$:

$$H(y/x) = \mu \log \mu - (1-\mu) \log(1-\mu).$$

Then the amount of information embedded in one symbol is given by

$$I_{Y \to X}^{(1)} = \{-r \log r - (1-r) \log(1-r)\} - \{-\mu \log \mu - (1-\mu) \log(1-\mu)\}$$
$$= [\eta(r) + \eta(1-r)] - [\eta(\mu) + \eta(1-\mu)],$$

with $\eta(x) = x \log x$ used for simplicity.

The amount of information per symbol is greatest for $r = \frac{1}{2}$:

$$\max I_{Y \to X}^{(1)} = 1 - [\eta(\mu) + \eta(1-\mu)].$$

To describe a channel in which information is lost completely, we set the maximal amount of information to zero:

$$\max I_{Y \to X}^{(1)} = 0.$$

Solving this yields:

$$\eta(\mu) + \eta(1-\mu) = 1, \tag{5.1}$$

so that

$$\mu = \frac{1}{2}.$$

Probability μ relates to time t as

$$\mu(\beta,t) = \left(1 - e^{-t/\beta}\right), \tag{5.2}$$

where β is the lead-time charactizing the rate of information degradation. For $t = 0$, the probability of a forecast distortion is zero, while for $t \to \infty$ it tends to unity, that is, to the probability of an actual event. The relationship between μ and β is complementary. If $\beta \to 0$ due to poor model quality, a forecast is unfaithful for practically any term. Conversely, for $\beta \to \infty$, its precision improves, and $\mu \to 0$. This means that the model is in good qualitative agreement with the empirical data.

The predictability horizon t_{hor}, that is, the utmost prediction range, is derived by setting $\mu = \frac{1}{2}$:

$$\mu(\beta, t) = \left(1 - e^{-\frac{1}{\beta}}\right) = \frac{1}{2},$$

The result is

$$t_{hor} = \beta \cdot \ln 2.$$

The probability of distortion can be written using t_{hor} as

$$\mu(\beta, t) = \mu\left(\frac{t_{hor}}{\ln 2}, t\right) = 1 - \exp\left(\frac{-t \ln 2}{t_{hor}}\right) \quad .$$

If β were time-independent, it would have been determined from stationary properties, for example, by expressing it in terms of the complexity of the predictand or in terms of the negentropy needed to prepare a forecast. For the above specific case of two contradictory events, t_{hor} is effectively independent of the quality of the prediction model. This is a universal postulate. Indeed, prediction is senseless beyond t_{hor} due to degraded precision. Thus, if we know the model used to make a prediction, we can determine t_{hor}.

It is only natural to aim at obtaining a model providing the maximum possible predictability horizon. If the relationship governed by a theory (a law) is known in outline, while the parameters (or constants) of the theory are unknown, the following method for deriving the parameters is proposed.

5.4 Optimization of Forecasting Methods

Suppose that a quantity $x_{t+\tau}$ is observed at discrete time instances $t + \tau$ in the past $(\tau = 0, -1, -2, \ldots, -n)$ to comprise a discrete time series $(x_t, x_{t-1}, x_{t-2}, \ldots, x_{t-n})$. Applying a polynomial prediction model presents $x_{t+\tau}$ as a sum of a non-random power series and random quantity ε_t:

$$x_{t+\tau} = \sum_{i}^{m} \frac{a_i}{i!}(t + \tau)^i + \varepsilon_{t+\tau} \quad . \tag{5.3}$$

Next, assume that $\varepsilon_{t+\tau}$ is a normal stationary random process with zero mean and correlation function $k_\varepsilon(\tau) = k_x(\tau)$ (a non-stationary process will be considered below). Then $x_{t+\tau}$ is also a normal stationary random process whose mean value is

$$\bar{x}_{t+\tau} = \sum_{i}^{m} \frac{a_i}{i!}(t+\tau)^i \ . \tag{5.4}$$

The task performed by forecasting is searching for coefficients a_i composing in aggregate the vector of parameters $\mathbf{a} = (a_1, a_2, ..., a_m)$. This problem is commonly solved by the method of least squares, so that the sum of the squares of the divergences of the experimental points from the smoothing curve is minimized:

$$\sum_{\tau=0}^{n}(x_{t+\tau} - \bar{x}_{t+\tau})^2 = \sum \varepsilon_{t+\tau}^2 = \min.$$

Given a Gaussian probability distribution for ε and using the principle of maximum likelihood (under the assumption that $\Sigma \varepsilon^2 = \min$) we can derive a reasonable number of coefficients a_i. Such an approach is generally recommended for obtaining agreement with historical data. We can also use other assumptions.

Let $\tau = 0,1,2,...,n$ be counts of future time instants. We consider two such instants, a running instant t and a relatively future instant $t+\tau$. What we are dealing with are two continuous systems of normally distributed random values x_t and $x_{t+\tau}$. It would be interesting to find out how much information about system $x_{t+\tau}$ can be acquired if system x_t is known to us.

Apparently, the state of $x_{t+\tau}$ can be predicted completely enough only if the amount of relevant information differs from zero (naturally, the prediction is made using only the past database x_t). The complete information exchange between the systems x_t and $x_{t+\tau}$ is written as

$$I_{x_t \leftarrow \to x_{t+\tau}} = \int\int_{-\infty}^{+\infty} f(x_t, x_{t+\tau}) \log \frac{f(x_t, x_{t+\tau})}{f(x_t)f(x_{t+\tau})} dx_t dx_{t+\tau} \ .$$

The complete information exchange $I_{x_t \leftarrow \to x_{t+\tau}}$ is a nonnegative value and vanishes only if x_t and $x_{t+\tau}$ are independent of each other. For x_t and $x_{t+\tau}$ to be independent when they are normally distributed, it is necessary and sufficient that the correlation moment be zero:

$$k(x_t, x_{t+\tau}) = 0 \ . \tag{5.5}$$

If τ is taken as fixed under the assumption that random x_t is stationary, equation (5.5) can be written as an equation producing the prediction range τ:

$$k(x_t, x_{t+\tau}) = k(t, t+\tau) = k_x(\tau) = 0.$$ (5.6)

In practice, of course, the function can only be derived for past time values ($\tau = 0, -1, -2, \ldots, -n$), but since the correlation function for stationary $k_x(\tau) = k_x(-\tau)$ is even, it is possible to employ the available information for $\tau > 0$. As a matter of fact, equation (5.6) provides a predictability horizon t_{hor} beyond which forecasting is impossible, since any information about x at instant $t+\tau+1$ vanishes (in the absence of periodic dependence).

The correlation function may be written as

$$k_x(t, t+\tau) = M[\tilde{x}(t)\tilde{x}(t+\tau)] = M(\varepsilon_t, \varepsilon_{t+\tau})$$
$$= M\{[x_t - \bar{x}(t)] \cdot [x_{t+\tau} - \bar{x}(t+\tau)]\},$$

or, allowing for equation (5.4), as

$$k(t, t+\tau) = M\left\{\left[x_t - \sum_i^m \frac{a_i}{i!}(t)^i\right] \cdot \left[x_{t+\tau} - \sum_i^m \frac{a_i}{i!}(t+\tau)^i\right]\right\}.$$ (5.7)

For a stationary process, the correlation function (which can be given by a regression equation) depends only upon τ and $\bar{a}$:

$$k_x(t, t+\tau, \bar{a}) = k_x(\tau, \bar{a}).$$

Setting it equal to a prescribed small quantity η (the extent of truthfulness of the forecast) yields an equation for interval τ_η:

$$k_x(\tau, \mathbf{a}) = \eta.$$ (5.8)

Solving (5.8) with respect to τ_η,

$$\tau_\eta = \varphi(\mathbf{a}, \eta),$$ (5.9)

yields an expression for the predictability horizon with respect to the vector $\bar{a}$ of a regression coefficient, and the truthfulness η of the forecast.

To define vector $\mathbf{a}$ corresponding to the utmost predictability horizon, equation (5.9) is written in the form

$$\tau_\eta = \varphi(\mathbf{a}, \eta) = \max.$$

Setting partial derivatives with respect to a_i to zero,

$$\left\{ \frac{\partial \varphi(\mathbf{a}, \eta)}{\partial a_i} \right\} = 0,$$

yields a_i values depending on η and previous values of $x_{t+\tau}$:

$$\mathbf{a} = f(\bar{x}_{t+\tau}, \eta), \quad \tau = 0, -1, -2, \ldots, -n.$$

In cases where only one realization of a random stationary process can be found over a fairly large time length T (under the assumption that the process is ergodic), the correlation function may be derived by averaging over time rather than over the set of observations. That is, equation (5.7) may take the following form:

$$k_x(\tau) = \frac{1}{T - \tau} \int_0^{T-\tau} x(t)x(t + \tau)dt.$$

If we restrict the number of power series members in equation (5.3) to two, then equation (5.8) relatively to τ can be carried to a numerical solution by defining a_1 and a_2 using the above procedure. The method discussed here is optimal for long term predictions.

5.5 Properties Shared by Prediction Methods

Assume that P is the predictive power of a given method, that is, the utmost amount of information about some unit of space derived for a specific prediction range that this method envisages. For example, if a method provides k elementary symbols of information per unit of space, then

$$P = k \max I,$$

where $\max I$ is the maximal amount of information available per symbol (it is always smaller than one; it might have equalled one in the absence of noise, which is impossible).

If a forecast is provided in the form of a random correlation function varying in space, then the power of the method can be written in the following form, which is similar to an expression familiar to information scientists:

$$P = \Omega \log_2\left(1 + \frac{S}{N}\right),$$

where Ω is a typical frequency (an *a priori* set characteristic of a class of prediction methods used when interpreting a prediction model as a Shannon communication channel – a kind of bandwidth), S is the mean power of the signal, and N is the mean power of the noise.

With reference to the quantity P, a fairly general statement may apply. Consider a method of prediction with power P used to predict some events with specific entropy h, which are to occur in the course of time. For $h > P$, prediction appears to be impossible; that is, one cannot predict events whose entropy is larger than the power of the prediction method. This statement is proved in much the same way as Shannon's second theorem in information theory.

Suppose that a message about an event is encoded by a predictability demon m using separate symbols so that it contains no excess information. To decode it without distortion, it is necessary that

$$k \geq \frac{m}{\max I},$$

where k is the number of elementary symbols per unit of space achievable with the method used. The problem concerning evaluation of prediction results can be solved using the following technique.

If the spectral density S_c of a message about a future event and the spectral density S_m of the prediction noise are known (in practice it suffices to know the ratio of their integral values), then the prediction spectral density $S_n(\Omega)$ can be found:

$$S_n(\Omega) = H^{-1}\left[\Omega \log\left(1 + \frac{\int_0^\infty S_c(\Omega)d\Omega}{\int_0^\infty S_m(\Omega)d\Omega}\right)\right], \tag{5.10}$$

where H^{-1} is the inverse function of $H = H[S(\Omega)]$, the system's entropy. The prediction's mean-square error σ^2 can be estimated by integrating equation (5.10):

$$\sigma^2 = \int\limits_0^\infty S_n(\Omega)d\Omega .$$

By way of example, consider the case of white noise, where function H^{-1} can be evaluated explicitly.

Suppose that prediction values x are normally distributed:

$$f(x) = \frac{1}{\sqrt{2\pi \cdot \sigma_n}} \cdot \exp\left(\frac{-x^2}{\sigma_n^2}\right).$$

Then the system's entropy is found from

$$H(x) = \log\left[\frac{\sqrt{2\pi \cdot e \cdot \sigma_n}}{\Delta x}\right]. \tag{5.11}$$

To model prediction noise, we use Gaussian white noise with constant spectral density $S_m = \text{const}$ and variance D_m, generally dependent on the time for which the prediction is made, but constant in spatial terms: $D_m = \sigma_m^2 = \sigma_m^2(t)$. The measurement made using the predictability demon is presented with a mean-square spread $\sigma_c = \text{const}$. From equations (5.10) and (5.11) we get

$$\log\left[\frac{\sqrt{2\pi \cdot e \cdot \sigma_n}}{\Delta x}\right] = \Omega \log\left[1 + \frac{\sigma_c^2}{\sigma_m^2(t)}\right].$$

The prediction's standard deviation is expressed as

$$\sigma_n = \frac{\Delta x}{\sqrt{2\pi e}}\left[\frac{\sigma_m^2(t) + \sigma_c^2}{\sigma_m^2(t)}\right]^\Omega ,$$

and the measure of the prediction accuracy $h = 1/\sigma_n \sqrt{2}$ is written as

$$h = \frac{\sqrt{\pi e}}{\Delta x}\left[\frac{\sigma_m^2(t)}{\sigma_m^2(t) + \sigma_c^2}\right]^\Omega ,$$

where Ω is the prediction method's bandwidth.

Expressing Δx in terms of Ω,

$$\Delta x = \frac{L}{2\Omega} ,$$

where L is the length of the conveyed function, we find

$$h = \frac{2\Omega\sqrt{\pi e}}{L}\left[\frac{\sigma_m(t)}{\sigma_m^2(t)+\sigma_c^2}\right]^{\Omega}.$$

Prediction involves determination of the constants in the laws followed by the predictand. Forecasts are valid only for those spatio-temporal regions in which the constants are determined. We can find the constants in many ways, but in each case we shall need information embedded in the prior database.

As an example, let us recall the Lorentz transformation, which is trivial from the standpoint of modern physics, but nevertheless embraces all the kinematic effects studied in the special theory of relativity. The transformation setup is suggested by the logic of averaged relationships displayed repeatedly in observed phenomena. Practically all the universal laws were formulated so as to conform to empirical data. The constancy of the quantity c (the speed of light) appearing in the Lorentz transformation was verified for a four-dimensional space in which $10^{-17}\,\text{cm} < \Delta x < 10^{28}\,\text{cm}$ and $10^{-27}\,\text{s} < \Delta t < 3\cdot 10^{17}\,\text{s}$. The constant was derived from repeated measurements and experiments, and is now defined as $2.99792458\cdot 10^{10}\,\text{cm}\,\text{s}^{-1}$.

At present, formalized estimates of constants appearing linearly in the law governing the variation of the predictand are commonly based on correlation and regression analysis.

As one chooses data to formulate a law governing the variation of the predictand, one may wonder whether there are any general rules to be taken into consideration.

Let us consider the time series of predictand x. As a rule, variations in x are strongly dependent on only a few factors, and weakly dependent on a host of minor factors. The aim of our investigation is to draw a demarcation line between the essential and the non-essential, that is, to ignore the minor factors and obtain the 'only possible' relation between the 'dependent' prediction and independent quantities that numerically express the causes. With this end accomplished, the constants entering into the prediction model can be determined by simultaneous substitution of dependent and independent quantities taken from a region with certain time and space coordinates. A quantity found in a region for which prediction is satisfactory can display the stability of the relationship between cause and effect.

The observed process can be regarded as the sum of the forecast's quasi-deterministic component and random noise. Instability in the relations between the effect and the causes (the latter including the minor factors) gives rise to instability in the characteristics of the random noise. For instance, it commonly amplifies data variance. The larger this variance, the less information about the true relation between cause and effect is conveyed.

5.6 The Connection Between Discounting and Non-stationarity

The empirical data one has at hand can be discounted. That is, they can be subject to a certain discrimination, with measurements made at some spatiotemporal interval. This impairs the accuracy with which constants in the model can be estimated and, consequently, the prediction range that can be achieved, since discounting brings about a reduction in the number of data points taken into consideration (let us recall that any reduction in the sample size causes dispersion of values formerly derived by averaging the sample data).

Therefore, if the information acquired by observation is thought of as a sum of true facts and misrepresentation, the relative share of the latter evidently grows with separation from the moment when a prediction is made in both the forward (future) and backward (past) directions. The rate of this growth (or reduction) depends upon the asymmetry of the function describing the way in which the dispersion of deviations from singular dependence varies, and upon whether or not this function can be extrapolated into the future. As experience shows, for shorter prediction times, closer time values should be used.

In order to predict some quantity, one has to determine its time dependence. One of the most developed mathematical methods to establish this dependence is smoothing of empirical data based on the principle of maximum plausibility. To apply this method, one needs a time series of the predictand's past values, that is, some information on its previous behavior.

Suppose that τ grows as one goes along the time axis into the past. Provided there is a series of n points on a plot depicting empirical data, one can construct a curve using these points, expressed in terms of a polynomial of degree $n-1$. However, this solution is hardly satisfactory, since the polynomial's coefficients are apt to be unstable. This instability is attributed to the fact that the variation in the behavior of a regular predictand under the action of stable factors has superimposed on it a host of comparatively weak random perturbations. Whatever probabilistic laws the individual perturbations might obey, the combined perturbations form what is practically a normal distribution. This property is made use of when the method of least squares is applied to predict behavior parameters.

Suppose that the true relation between x and τ is expressed as $x = \varphi(\tau)$. Consider the value of argument τ for the case where the value of x is x_τ. By virtue of the causes specified above, the random x_τ practically follows a normal distribution with expected value $\bar{x}(\tau)$ and mean square deviation σ_τ. The normal law of the x_τ distribution can be written as

$$f_\tau = \frac{1}{\sigma_\tau \sqrt{2\pi}} \cdot \exp\left(\frac{-[x_\tau - \bar{x}(\tau)]^2}{2\sigma_\tau^2}\right).$$

We now establish the probability that the system of generally dependent (or at least correlated) random values $x_1, x_2, \ldots, x_n$ assumes values falling in the range $(x_\tau, x_\tau + dx_\tau)$ (where $\tau = 1, 2, \ldots, n$). This probability has been estimated with the help of the normal distribution law in n-dimensional space:

$$f(x_1, x_2, \ldots, x_n) dx_1 dx_2 \ldots dx_n =$$

$$= \frac{\sqrt{[c]}}{\sqrt{2\pi}} \cdot \exp\left\{-\frac{1}{2}\sum_{\tau=0}^{n}\sum_{t=0}^{n} C_{t\tau} \tilde{x}_\tau \tilde{x}_t\right\} dx_1 dx_2 \ldots dx_n, \tag{5.12}$$

where $\tilde{x}_t = x_t - \bar{x}(t)$, $C = \|C_{t\tau}\|$ is the inverse matrix of the correlation matrix K, and $[c]$ is its determinant. Assuming that $K = \|K_{t\tau}\|$ gives

$$C_{t\tau} = (-1)^{\tau-t}\frac{M_{t\tau}}{[K]} = (-1)^{\tau-t} M_{t\tau} \cdot [c],$$

where $[K]$ is the determinant of the correlation function and $M_{t\tau}$ its minor, obtained by deleting the τ-th row and the t-th column from $[K]$. The correlation function can be written in outline as

$$\|K\| = \begin{Vmatrix} K_{11} & K_{12} & \cdots & K_{1n} \\ K_{21} & K_{22} & \cdots & K_{2n} \\ \cdots & \cdots & \cdots & \cdots \\ K_{n1} & K_{n2} & \cdots & K_{nn} \end{Vmatrix} ,$$

where $K_{\tau t} = M(\tilde{x}_\tau \tilde{x}_t)$ are correlation moments characterizing the pairwise correlation of values $\tilde{x}_\tau = x_t - \tilde{x}_t$ entering into the system. We can rewrite $\|K\|$ as

$$\|K\| = \begin{Vmatrix} r_{11}\sigma_1\sigma_1 & r_{12}\sigma_2\sigma_1 & \cdots & r_{1n}\sigma_n\sigma_1 \\ r_{21}\sigma_1\sigma_2 & r_{22}\sigma_2\sigma_2 & \cdots & r_{2n}\sigma_n\sigma_2 \\ \cdots & \cdots & \cdots & \cdots \\ r_{n1}\sigma_1\sigma_n & r_{n2}\sigma_2\sigma_n & \cdots & r_{nn}\sigma_n\sigma_n \end{Vmatrix} .$$

Using the principle of maximum plausibility, let us pick mean values $\bar{x}(1), \bar{x}(2), \ldots, \bar{x}(n)$ so as to maximize expression (5.12). This task is accomplished by maximizing

$$\sqrt{[c]} \exp\left\{ -\frac{1}{2} \sum_{\tau=0}^{n} \sum_{t=0}^{n} C_{\tau t} \tilde{x}_\tau \tilde{x}_t \right\} = \max . \tag{5.13}$$

From (5.13) the requirement follows that

$$\sum_{\tau=0}^{n} \sum_{t=0}^{n} C_{\tau t} \tilde{x}_\tau \tilde{x}_t = \min . \tag{5.14}$$

If values $\tilde{x}_\tau$ and $\tilde{x}_t$ are uncorrelated for $\tau \neq t$,

$$K_{\tau t} = M(\tilde{x}_\tau \tilde{x}_t) = 0 ,$$

then equation (5.14) takes the form

$$\sum_{\tau=0}^{T} \frac{\tilde{x}_\tau^2}{\sigma_\tau^2} = \min . \tag{5.15}$$

For a specific case, where random x_τ is stationary, i.e., all σ_τ are equal, requirement (5.15) is in line with the traditional least squares method:

$$\sum_{\tau=0}^{T} \tilde{x}_\tau^2 = \min . \tag{5.16}$$

By using in a certain way the information fixed in a single observation, it is possible to determine a confidence interval for the true value of $x(\tau)$ at a predetermined confidence probability. Assuming that the amount of information provided by a single observation is independent of observation time, equation (5.16) is legitimate. However, this assumption is not quite appropriate here, for the amount of information coming from observations in the recent past differs from that coming from the remote past. In other words, with the course of time, information becomes depreciated, or discounted. Many workers have accounted for this depreciation through equation (5.16) using discounting weight factors W_τ that set up a correspondence between observations and the times at which they were made:

$$\sum_{\tau=0}^{T} W_\tau \tilde{x}_\tau^2 = \min. \tag{5.17}$$

We now try to establish a relationship between the weight function and information depreciation. To do this we need to determine the amount of information contained in observation $\tilde{x}_\tau$.

In our case, the entropy of observation x_τ at time τ is written as

$$H_\tau = \log \frac{\sqrt{2\pi e}\sigma_\tau}{\Delta x}$$

while the entropy change ΔH over time $\Delta\tau$ is

$$\Delta H_\tau = \log \frac{\sigma_\tau + \Delta\tau}{\sigma_\tau} = \log \frac{\sigma_\tau + \Delta\sigma_\tau}{\sigma_\tau}.$$

Given a stationary process, $\Delta H_\tau = 0$, since $\Delta\sigma_\tau = 0$. That is, under the assumption that the dispersion of empirical data about some mean $\bar{x}(\tau)$ is time-independent, no discounting of information takes place, and the weight function is identically equal to one. Proceeding from this, we can state the following.

Instead of searching for the expected value of an arbitrary nonstationary process with uncorrelated deviations by the least squares method, one can derive the expected value of such a stationary process by the least weighted squares method, with weight coefficients estimated according to

$$W_\tau = \left(\frac{\sigma_0}{\sigma_\tau}\right)^2.$$

The reverse statement also holds true: introduction of discounting weights into the least squares method is tantamount to the hypothesis that a time series of empirical data is non-stationary such that

$$\sigma_\tau^2 = \frac{\sigma_0^2}{W_\tau}.$$

The validity of these statements may be seen by comparing equations (5.15) and (5.17). If the entropy of observation x_τ is $H_\tau = C_1 \log \sigma_\tau$ and $\sigma_\tau = \sigma_0 \cdot W_\tau^{-\frac{1}{2}}$, then we get $H_\tau = C - \frac{1}{2}\log W_\tau$, or $W_\tau = 2^{2(C-H_\tau)}$.

It should be emphasized that for the entropy of x_τ it follows from $H_\tau = C + \frac{1}{2}\log \sigma_\tau$ that

$$H_\tau = C_1 + \tfrac{1}{2}\log \sigma_\tau - \tfrac{1}{4}\log W_\tau = C - \tfrac{1}{4}\log W_\tau.$$

This yields a convenient expression for estimating the discounting factor W_τ:

$$W_\tau = 2^{4(C-H_\tau)}.$$

5.7 Conclusion

It is admitted in the present-day literature that there are many gaps in the theory of prediction reliability evaluation. The difficulties in tackling this problem stem from limits in the strength of the information-theoretic semantic and statistical approaches as applied to interpreting the prediction process. In this chapter we have tried to extend (Shannon's) information-theoretic approach in which prediction is looked upon as signal transmission via a noisy transmission channel. We have thereby derived a number of helpful expressions that fairly rigorously describe the most general properties of prediction methods and models. Further extension of the theoretical results obtained will probably take two separate lines. Firstly, it is important to relate the entropic character of a prediction model (that is, of a theory adequate to some extent to the predicted event or quantity) to the theory of subjective probabilities. This will significantly extend the area of interpretation of the results obtained.

Secondly, a prediction model should be constructed using the Bayesian approach. Previous attempts to do so were not connected with the information-theoretic approach discussed here.

Acknowledgements

I am deeply grateful to the editor, Professor Yu. A. Kravtsov, whose salutory criticism has greatly improved the presentation of this chapter, and to my co-workers for helpful discussions.

References

5.1   V.A. Lisichkin: Prognostics. Terminology. Moscow: Nauka, 1990. p. 47 (in Russian)

5.2   J. Zeman: Knowledge and information. Moscow: Nauka, 1965. p. 320 (in Russian)

5.3   K.E. Morozov: Philosophical aspects of information theory. In: The natural science philosophy. Moscow: Znanie, 1966. p. 451 (in Russian)

5.4   W.R. Ashby: An introduction to cybernetics. London: Chapman and Hall, 1956

5.5   A.A. Kharkevich: Problems of cybernetics 4. Moscow: Nauka, 1960. p. 195 (in Russian)

5.6   A.A. Kharkevich: Essays on the general theory of systems. Moscow: Nauka, 1965. p. 428 (in Russian)

5.7   W. Miles: The measurements of the value of scientific information. In: Operation research in research and development. Proc. Conf. at Case Institute of Technology. New York, London: John Wiley and Sons, 1963

5.8   R. Ackoff, P. Emery: On purposeful systems. Chicago, New York: Aldine-Atherton, 1972

5.9   C. Shannon: A mathematical theory of communication. Bell System Techn. J. 27(3), 379–423 (1948); 27(4), 623–656 (1948)

5.10 T.M. Reed, K.E. Gubbins: Applied statistical mechanics. New York, 1973. p. 271

5.11 D. Kenyon, G. Steinman: Biochemical predestination. New York: McGraw-Hill, 1969

6. Prediction of Time Series

M. A. Sadovskii and V. F. Pisarenko

6.1 The Problem

A classic statement of the problem of predicting stationary time series $x(t)$ is as follows [6.1, 6.2]. Suppose that a stationary random time series $x(t)$ is defined on time axis $t \in [-\infty, +\infty]$. To simplify the discussion, let us assume that the mean value of the process is zero:

$$E\, x(t) = 0.$$

Knowing how $x(t)$ is realized up to the instant $t = 0$, we seek to forecast values $x(\tau)$ of this process in the future ($\tau > 0$), provided the optimal forecast $\hat{x}(\tau)$ by definition minimizes the prediction's mean-square error

$$E\,|x(\tau) - \hat{x}(\tau)|^2 = \min.$$

In general terms, the solution of this problem is known: the optimal forecast appears as

$$\hat{x}(\tau) = E\left[x(\tau)|x(t), t \le 0\right], \tag{6.1}$$

where the right-hand term denotes conditional expectation of the random value $x(\tau)$ under the condition that the realized values of $x(t)$, for $t \le 0$, are given.

This formulation of the problem naturally led to numerous generalizations, and new problems were formulated, such as the prediction of multidimensional processes and fields, the filtration, interpolation, and prediction of generalized processes, and so on. For the purposes of this chapter, the most essential point is that the stochasticity of whatever physics might be involved is assumed to be given *a priori*.

Now, what are the mechanisms whereby quantities and processes habitually looked upon as random occur in practice?

6.2 Genesis of Random Phenomena

As far as the problem under discussion is concerned, three classes of random phenomena (quantities) can be distinguished as follows [6.3].

(1) Quantities η_n arising from interaction of summands ξ_i which are independent (or weakly interdependent) and of the same order of magnitude:

$$\eta_n = \xi_1 + \xi_2 + \ldots + \xi_n . \tag{6.2}$$

Some of the ξ_i summands may be of a non-random nature. But irrespective of their nature, quantity η_n behaves like a random Gaussian variable, in accordance with the central limit theorem of probability theory. Examples of such random variables include the stress field within a loaded rock sample pierced through with micro-fractures or the seismic field within a medium with density or elastic-constant inhomogeneities.

Note that in these cases it is possible (at least in principle) to predict a field's value by measuring all stochasticity factors (fractures and inhomogeneities) individually. However, such measurements are impractical from both the technical and the computational points of view, and it is more convenient to treat the field as a random one.

The above random variables can be called phenomena with many independent factors in linear superposition, so many that it is practically impossible to identify individually all the factors at work (recall, for example, the Brownian motion of a microparticle in a liquid, or electronic shot noise).

(2) Phenomena or processes generated by unstable nonlinear mapping or dynamical systems [6.4, 6.5]. Mapping produces a sequence of quantities x_n (which, generally speaking, are vectors):

$$x_n = F(x_{n-1}; \lambda), \quad n = 1, 2, \ldots, \tag{6.3}$$

where F is a nonlinear function, λ a parameter, and x_0 the arbitrarily given initial value. In the case of continuous-time dynamical systems, the $x(t)$ function (presumably a vector function) is a solution of the nonlinear differential equation

$$F\left(\frac{d}{dt}, x(t); \lambda\right) = 0, \tag{6.4}$$

where the coefficients are λ-dependent and the initial conditions for $x(t)$ are given at $t = 0$. For some F, sequences x look like random processes, although they are solutions to non-random differential equations. This phenomenon is called 'dynamical chaos' [6.4, 6.5]. Such chaotic phenomena are quite common in both nature and technology. For example, practically all computer algorithms for generating random numbers are of the form (6.3) and related to dynamical chaos.

The number of degrees of freedom can also be determined for random events of the second type. For convenience, we can discuss dynamical systems in terms of what we call the phase space. As an example, consider an m-dimensional space determined by values $x(t)$, $x'(t),...,x^{(m-1)}(t)$, where m is the order of the differential equation. A dynamical system trajectory passes through each point of the phase space (with the exception of some special points). The evolution of such a system is portrayed by a curve in a phase space. Thus, we can use m to denote the number of degrees of freedom for a chaotic dynamical curve. Sometimes, as $t \rightarrow \infty$, the curve trajectory becomes captured by some lower-dimensional ($d < m$) 'attractor' instead of occupying the whole space. An attractor with a non-integer dimension d is called a strange attractor [6.4].

As a matter of fact, randomness in chaotic dynamics is characterized by a finite number of independent factors (the number can sometimes be as large as several tens or hundreds or more).

Dynamical chaos occurs under certain circumstances in turbulent fluid flows and magnetohydrodynamic fluxes, during convection in viscous fluids, and so on.

(3) Random quantum phenomena. These include, for example, radioactive decay and electron diffraction. Their randomness is probably the most perfect (unpredictable) among the known types of randomness, as its nature is absolutely undetermined. However, we only mention this source of randomness for completeness in this study.

A number of new features in formulating the time series prediction problem have recently appeared in the literature. They stem from (a) dynamical chaos theory, (b) new applications of the point processes prediction technique (specifically, to predict earthquakes [6.6, 6.7]), and (c) a new view of the nature of factors setting limits to predictability [6.8].

6.3 Time Series Prediction Based on Dynamical Chaos Theory

In the past few years, some efficient new methods for predicting time
series produced by dynamical systems have been developed [6.9, 6.10]. It
is important that in using these methods the system itself, i.e., function F
in equations (6.3) and (6.4), is not assumed to be known, but is reconstructed (with some error naturally introduced) using a finite realization
of $x(t)$. In fact, the prediction problem is divided into two parts:
reconstruction of the phase space using the realization of $x(t)$ and
interpolation, using appropriate methods, of the forecast for each
individual point of the phase space (or, to be more exact, for each point's
neighborhood). The first part of the problem is solved by means of the
Packard–Takens theorem [6.11, 6.12], which shows how to construct a
finite-dimensional space incorporating the system's phase space using
only one realization of $x(t)$. Specifically, reconstruction of the phase
space using the Packard–Takens theorem yields an embedding space
containing the true phase space.

For the second part, we can employ methods of multidimensional
space interpolation. The result is that even without knowing equations for
the dynamical system $F(x;\lambda)$ or $F(d/dt;\lambda;x)$ we can obtain a fairly
close approximation $\hat{F}(x;\lambda)$ or $\hat{F}(d/dt;\lambda;x)$, which provides an acceptable forecast for a limited time interval on the basis of equation (6.4).

In many cases, this forecasting method using estimates $\hat{F}(x;\lambda)$ or
$\hat{F}(d/dt;\lambda;x)$ is far more efficient than conventional linear prediction
using the correlation function $B_T(\tau)$ of the $x(t)$ process:

$$B_T(\tau) = \frac{1}{T} \int_{-T}^{0} x(u)x(u-\tau)\,du.$$

An example considered in [6.13] shows that in the case of convection
of a viscous liquid heated from below the nonlinear dynamical prediction
method is 50 times as good (in terms of the mean-square error) as
methods based on a standard linear prediction model. Very impressive
results are also described in [6.9, 6.10] for predicting a dynamical system
unknown to the observer. Examples of successful prediction based on
reconstructing a strange attractor along with its limit invariant measure
using as few as 75, 50, 25, and even 10 points (!) are given in [6.9]. Of
course, nonlinear prediction is not always better than optimal linear

prediction, but at any rate it is worth doing if only to compare the results with those of the linear method.

As a matter of fact, nonlinear prediction involving interpolation of trajectories in phase space resembles weather forecasting in the pre-computer era. In those days, a meteorologist trying to forecast (say) pressure at height 1000 m would have to search through isobar charts for earlier times and find the isobar pattern most similar to the pattern for that day. Then by looking at the earlier pattern for the following day he could predict the next day's pressure. In this procedure an isobar chart – a two-dimensional drawing – played the part of a phase space. The special merit of computers is that they offer the possibility of utilizing spaces of more than two dimensions as well as a variety of elaborate methods for multidimensional interpolation.

6.4 Prediction of Point Processes

Considerable advances have been made in the area of point process prediction [6.6, 6.7]. For this kind of prediction, one uses observations of a past process $y(t)$, $t \leq 0$, to forecast future time moments $t_1, t_2, \ldots$ of a point process $x(t)$ connected with the process $y(t)$. Roughly speaking, point process $x(t)$ is given by a sum

$$x(t) = \sum_i \delta(t - t_i),$$

with all the information contained in time moments t_i. For example, in one of the most important applications, t_i are moments at which powerful earthquakes occur in a given region.

Values $\{t_i\}$, $i = 1, 2, \ldots$, can be predicted in different ways. For example, we can predict the time t_1 at which a point process closest to $t = 0$ is going to happen, then, at time $t = t_1$, predict the next process t_2 and so on. Or we can use another method, namely, to predict the occurrence or non-occurrence of event t_1 in the nearest time interval Δt. This method is employed, for example, in [6.6], where a point process $t_1, t_2, \ldots$ gives the times of powerful earthquakes in a given region. Errors of two kinds can be introduced using this method, 'target missing' errors (failures to predict an event occurring within the Δt interval) and 'false alarm' errors (prediction of events that do not in fact occur in the Δt interval). For stationary processes, such errors can be characterized in terms of time-independent probabilities α, β. To formulate the problem of optimal prediction technique, we must introduce the *loss function* $\gamma(\alpha, \beta)$. By definition, an optimal forecast is one minimizing the *mean loss function value*.

Apparently, this kind of prediction poses a much more intricate problem than forecasting a stationary random process, where a minimal mean-square error serves as an optimal forecast criterion. Although the mean-square error indicates the prediction situation on average, a forecast that minimizes the mean loss function $\gamma(\alpha, \beta)$ is in general different from a 'mean-square-optimal' prediction and can respond better to practical needs reflected by the choice of loss function $\gamma(\alpha, \beta)$.

The above approach can be generalized by introducing new characteristics for prediction quality. For example, one can try to predict three versions of situation development – no event, one event, and more than one event – within time interval $(t, t + \Delta t)$. Applied to earthquake forecasting, this corresponds to predicting separately a single seismic event and two or more multiple quakes of about the same magnitude. Such individual behavior may be important in practice, since prediction of powerful shocks that may follow in series is a matter of vital significance. It may be recalled that paired shocks of about the same strength are not infrequent in seismological experience. In the Gazli region of Uzbekistan in 1976, for example, two strong quakes with a magnitude of about 7.2 occurred within an interval of 40 days.

Timeliness might also be used as an additional characteristic of prediction quality. However, introduction of additional characteristics leads to new sets of errors or to quantitative changes in other prediction characteristics. To prepare an optimal forecast, it is eventually necessary to introduce a loss function of the prediction's average characteristics or probabilities of errors of different origin and to find the method of prediction that minimizes this loss function.

6.5 The Nature of Errors Hindering Prediction

In a classic prediction problem, the error is characterized by only one value, namely, its standard deviation (for a fixed prediction time τ). This is a rather rough and uninformative way to describe prediction accuracy. A much more detailed and specific description can be obtained if we know the conditional local mean-square prediction error, i.e., the error for the neighborhood of any point in the phase space. Such information can in principle be obtained when a dynamic forecast involving an unknown function F is made by means of phase space interpolation [6.9, 6.10. 6.13]. The ordinary mean-square error equals a conditional local error averaged over the entire phase space (with respect to the natural invariant measure). Such further specification of the mean-square error provides additional information that is useful for the prediction problem.

Another procedure for increasing the informativeness of mean-square errors is proposed in [6.8] and applied to dynamical systems forecasting. It involves partitioning the noise contained in a dynamical system into three groups according to its origin:

- Noise of the measurement type, which can in principle be lowered by reducing instrument noise and by taking more frequent samples from a continuous process;
- Noise due to the inaccuracy of the dynamical system model used to make the prediction, which can in principle be reduced by complicating the model and more accurately fitting it to the actual dynamical system;
- Noise that cannot be reduced however hard we try ("the whole universe influencing our dynamical system" [6.8]).

Once we know the relative strength of these sources of prediction error we can, generally speaking, define the one that most strongly affects prediction accuracy. For example, if these three sources enter additively into a dynamical system, the variance of their sum is dominated by the largest summand's variance. Therefore, to improve forecasting we need to reduce the noise with the largest variance.

The noise of the third type that cannot be reduced using present-day technology can actually define the limits of effective forecasting – the 'predictability horizon' [6.8].

To generalize this approach, we suggest decomposing the cumulative noise ξ in a dynamical system into a number of components $\xi_1,...,\xi_m$. In the simplest case of additive noise,

$$\xi = \xi_1 + ... + \xi_m$$

and ξ_i are independent, while the variances σ_i^2 of the components are additive as well:

$$\sigma_\xi^2 = \sigma_1^2 + \ldots + \sigma_m^2 \, . \tag{6.5}$$

In the general case, components $\xi_1, \ldots, \xi_m$ may appear in different parts of the dynamical system and exert different influences on the quality of forecasts.

It is expedient to subdivide noise into components according to the costs involved in reducing a given noise. The costs can be given by a set of monotonically decreasing functions $\varphi_i(\mu)$ specifying a $1/\mu$-fold reduction cost for the i-type noise variance σ_i^2. In case σ_k^2 reduction is impracticable, one can formally set

$$\varphi_k(\mu) = \infty, \quad \mu < 1.$$

The total cost of a reduction of the i-type noise variance is given as

$$\varphi = \varphi_1(\mu_1) + \ldots + \varphi_m(\mu_m) \, . \tag{6.6}$$

With equation (6.5) taken for simplicity, the expression for the resulting variance is:

$$\sigma_\xi^2 = \mu_1 \sigma_1^2 + \ldots + \mu_m \sigma_m^2 \, . \tag{6.7}$$

It is only natural to pick values $\mu_1, \ldots, \mu_m$ such that at fixed costs φ the combined variance σ_ξ^2 can be minimized. Consider an example where

$$\varphi_i(x) = \frac{1}{x^{\alpha_i}} - 1; \quad i = 1, \ldots, m,$$

with α_i preset as positive constants. In this case, optimal 'weights' of noises μ_i minimizing the cumulative noise (6.7) at given costs (6.6) are given by:

$$\mu_i = \left(\frac{\lambda_0 \alpha_i}{\sigma_i^2} \right)^{\frac{1}{1+\alpha_i}}, \tag{6.8}$$

where λ_0 is the root of the equation

$$\lambda^{\frac{1}{1+\alpha_1}} \left(\frac{\alpha_1}{\sigma_1^2} \right)^{\frac{1}{1+\alpha_1}} + \ldots + \lambda^{\frac{1}{1+\alpha_m}} \left(\frac{\alpha_m}{\sigma_m^2} \right)^{\frac{1}{1+\alpha_m}} = \varphi + m \, .$$

Thus it is expedient to distribute the costs φ of reducing σ_ξ^2 variance over components with weights μ_i determined from equation (6.8).

Decomposition of noise using the suggested procedure seems reasonable, for even if no noise reduction is actually achieved, its future prospect is defined, and that is a start.

6.6 Prediction of Strong Earthquakes

In this concluding section we shall consider the prospects for applying the above new approaches to forecasting powerful earthquakes. The method described in Sect. 6.4 is used for these purposes [6.6, 6.7]. The 'target missing' and 'false alarm' errors mentioned above exemplify the variety of errors that forecasting involves.

Minimization of the loss function of these errors determines optimal prediction. The number of conditional probabilities can in principle be larger than two, allowing a better representation of the optimal prediction. The technique of partitioning noise into three components suggested in [6.8], as well as its generalization described in Sect. 6.5, can apparently be useful for quake forecasting and other practical purposes. It actually provides insight into the nature of noise and indicates the cheapest way to reduce the noise encountered in practice.

The most perplexing question is whether our knowledge about dynamical chaos and nonlinear prediction is applicable to predicting earthquakes. There are reasons for us to be sceptical about such a possibility. First of all, unlike the cases of convection, hydrodynamics or magnetodynamics, there are no equations specified for the earthquake process. While problems related to the other cases have solutions (at least in some specific instances) that are proven theoretically, or numerically, to possess the properties of dynamical chaos within a small number of degrees of freedom, it is still unclear for what physical field the earthquake equation is to be derived. Quakes occur within the fractured block structure of the lithosphere, the Earth's inhomogeneous rocky outer layer, as a result of tectonic deformations. It seems reasonable to assume that seismic behavior is strongly influenced by various physical and chemical processes, such as those related to heat, electricity, fluid flows along fractures, and so on.

Is it really possible to derive an adequate equation for a stress field within such a complex medium? Besides, what does the field of forces look like in such circumstances? The answers to these questions are not at all simple.

Furthermore, it should be pointed out that the well-developed forecasting methods [6.9, 6.10, 6.13] can appear inefficient where motions with dynamical chaos properties described by nonlinear equations are concerned, owing to the great number of degrees of freedom. If this number is greater than 10, the amount of observational data needed to estimate forecasting dynamics and prognostic techniques may be prohibitive for many practical problems. There is no justification as yet for counting on the discovery of low-dimensional systems controlling quakes.

Another complication in quake prediction is the high noise level involved. In the successful cases of dynamical forecasting mentioned above, the noise level is 1–2 orders of magnitude lower than that of the genuine signal (the field), but in the cases involved in an earthquake (for example, recording a stress field), the noise and the true signal are comparable in level. It should be readily apparent that even with a dynamical system most favorable to forecasting, the comparability of exogenous noise and signal make any forecast unreliable.

Given all this, are there any prospects for applying the latest findings concerning dynamical chaos to the problems of quake forecasting? First and foremost, it should be stated that as a result of the dedicated efforts of seismologists it has become possible to specify some more or less reliable quake precursors (acting in a statistical rather than a deterministic manner) and to develop methods of long-term forecasting (1–3 years in advance). Although these are not yet as simple and reliable as one would like, they are good enough to diminish losses due to quake disasters [6.14]. Their statistical significance is indubitable. Moreover, there have been cases where they enabled successful short-term prediction [6.15]. These facts suggest that the most reliable precursors might be used to reconstruct the seismic process phase space. Once we have assembled precursors which in combination indicate faithfully enough that a powerful quake is going to happen, we can go on to employ the nonlinear dynamics techniques of trajectory interpolation in phase space.

However, it seems to us that the most promising approach for this application is discretization of a seismic process in both time and space. In fact, discretization is realized in the above long-term forecasting method. Naturally, forecasts for 1–3 years ahead and for one day ahead are incommensurable in terms of uncertainty. However, the longer the term the better the reliability. It should be pointed out that phase space discretization may significantly reduce chaos and even give rise to 'deterministic behavior' described by what is called symbolic dynamics [6.16, 6.17]. The question of how to reconstruct the phase space in order to obtain the best possible predictions is interesting and intriguing, but has not been approached so far.

We can only make cautious judgements about the possibility of reliably predicting earthquakes. Indeed, any equation or mathematically proven statement suggesting that forecasting is impossible is always based on assumptions or simplifications regarding natural seismic processes. For instance, sometimes certain phenomena are treated in terms of concepts adopted from destruction mechanics, implying that no satisfactory forecasts are possible under given conditions. But apart from phenomena covered by destruction mechanics, there are plenty of other factors (of electrical, chemical, magnetic, or other origin) that have to be accounted for in real life. Besides, destruction mechanics deals with a limited frequency range, while useful prognostic information can be contained in infrasonic or ultrasonic oscillations outside the range, which only animals or special instruments can detect. It is therefore premature to conclude that it is impossible to predict earthquakes with some given accuracy. We can only speak about the limited forecasting capabilities of individual methods in view of their simplifying assumptions.

References

6.1 A.N. Kolmogorov: Interpolation and extrapolation of stationary random sequences. Izv. Acad. Sci. USSR, Mathematical series 5(1) 3–14 (1941) (in Russian)

6.2 N. Wiener: Extrapolation, interpolation and smoothing of stationary time series. Cambridge, MA: MIT Press, 1949

6.3 M.A. Sadovskii, V.F. Pisarenko: Randomness and instability in geophysical processes. Izv. Acad. Sci. USSR, Earth physics series 2, 3–12 (1989) (in Russian)

6.4 G. Shuster: Deterministic chaos. Weinheim: Physik-Verlag, 1984

6.5 Yu.I. Neumark, P.S. Landa: Stochastic and chaotic oscillations. Moscow: Nauka, 1987 (in Russian)

6.6 G.M. Molchan: Strategies in predicting powerful earthquakes. In: Computer-aided analysis of geophysical fields, pp. 3–27. Moscow: Nauka, 1990 (in Russian)

6.7 G.M. Molchan: Structure of optimal strategies in earthquake prediction. Physics of the Earth and Planetary Interiors 61(1) 84–98 (1990)

6.8 Yu.A. Kravtsov: Randomness, determinateness and predictability. Sov. Phys.-Uspekhi 32(5) 434–445 (1989)

6.9   M. Casdagli: Nonlinear prediction of chaotic time series. Physica D 5(2) 335–336 (1989)

6.10 H. Abarbanel, R. Brown, B. Kadtke: Prediction in chaotic nonlinear systems: methods for time series with broadband Fourier spectra. Phys. Rev. 41(4) 1782–1807 (1990)

6.11 N. Packard, J. Crutchfield, J. Farmer, R. Shaw: Geometry from a time series. Phys. Rev. Lett. 45(9) 712–716 (1980)

6.12 F. Takens: Detecting strange attractors in turbulence. In: Lect. Notes in Math. 898, pp. 366–381. Berlin: Springer-Verlag, 1981

6.13 J. Farmer, J. Sidorowich: Predicting chaotic time series. Phys. Rev. Lett. 59(8) 845–848 (1989)

6.14 A.M. Gabrielov, O.D. Dmitrieva, B.I. Keilis-Borok et al.: Long-term earthquake prediction: methodical recommendations, pp. 2–127. Moscow: Nauka, 1986 (in Russian)

6.15 K. Mogi: Earthquake prediction. Tokyo: Academic Press, 1985

6.16 R. Dowen: Equilibrium states and the ergodic theory of Anosov diffeomorphisms. Lecture Notes in Math. 470, p. 280. Berlin: Springer-Verlag, 1975

6.17 A.A. Samarskii (ed.): Computers and nonlinear phenomena: information science and modern natural history. Moscow: Nauka, 1988 (in Russian)

7. Fundamental and Practical Limits of Predictability

Yu. A. Kravtsov

7.1 Predictability

The tremendous intellectual breakthrough achieved in the twentieth century, based on the entire previous development of knowledge, cannot but create the illusion that science is almighty and capable of tackling any technical, medical, or other problem, however intricate it might be. To the present day, science continues to be a fundamental factor of progress. However, there is one issue where progress has clearly slowed down and where science fails to manifest its potency despite increases in the speed, precision, and information-handling capacity of modern instrumentation. What I mean is the problem of predictability. What bounds forecasting terms? Why don't state-of-the-art computing facilities always suffice to outline even the nearest future? What exactly is lacking in modern scientific techniques that would provide more efficient predictions? Why are improvements in our forecasting skill far from proportional with our investment?

In this review of the issue of predictability, we shall consider the relatively simple case of dynamical processes involving oscillations that obey a more or less complex law. By way of this example, factors limiting predictability can be elucidated. I can say in advance that they are rather numerous, so we shall restrict our discussion to basic, if not always apparent, factors.

So what are the limits to predictability?

7.2 Real, Observed, and Model Processes

Each time we take up forecasting we start by analyzing an observed process $y(t)$ that we assume to be an event tracked by our instrumentation. However, $y(t)$ is far from identical to the real physical process $x(t)$, since observation or measurement is invariably subject to distortion.

To begin with, instruments commonly filter signals, thereby introducing certain spectral distortions, for example, an ordinary time lag, $y(t) = x(t - \tau)$, which is typical of many systems in technology, economics, meteorology, and so on. A strong signal or an inadequate instrument can give rise to nonlinear distortions. Further, there is noise $v(t)$ due to measurement, called instrument noise for simplicity, which plays a principal role. It prevents estimation and sometimes detection of weak signals.

Finally, there is a vector of observation $y(t)$ whose dimension q_y may be smaller than the dimension q_x of the real process, since the number of measured parameters is always smaller than the total number of degrees of freedom involved in the motion. These sorts of distortion can be called dimensional distortions.

In the simplest case where spectral, dimensional, and nonlinear distortions are negligible, and the noise has an additive character, observation $y(t)$ is related to the real process $x(t)$ by

$$y(t) = x(t) + v(t). \tag{7.1}$$

A forecast is always founded upon the representations provided by a model and is the result of a simplified – approximated or idealized – description of a real system. The differential equation governing a real dynamical system can be written symbolically as

$$M_x\left(\frac{d}{dt}, x; \alpha_k, f_j\right) = 0. \tag{7.2}$$

Here M_x is the system's operator, d/dt is the differential operator, x is the set of variables that are essential for the chosen level of precision in the description of the problem at hand, α_k are the real system parameters, and $f_j(t)$ are all inessential, minor, and fluctuational factors. The division of variables into essential and inessential involves some arbitrariness. Depending on how it is handled, we obtain one or another approximated model equation

$$M_z\left(\frac{d}{dt}, z; a_k, 0\right) = 0, \tag{7.3}$$

where a_k are parameter estimates (for the idealized case, $a_k = \alpha_k$), and symbol '0' is placed where inessential factors $f_j(t)$ appeared in the real

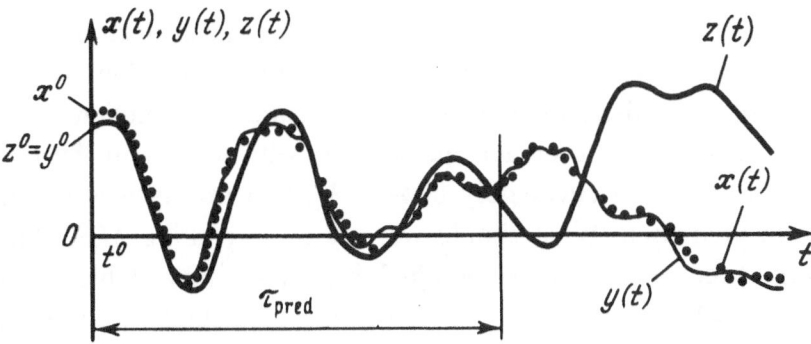

Fig. 7.1. Processes considered in making predictions: the real physical process $x(t)$ (dotted line), the observed process $y(t)$ (thin line), and the process $z(t)$ predicted by the model (heavy line)

state equation (7.2). Neglect of these factors is eventually what makes idealization.

In tackling the problem of predictability we are actually dealing with two processes, the observed one $y(t)$ and the model one $z(t)$, while the real process $x(t)$ remains as it were in the background; it is neither directly followed nor does it have a definite dimension nor is it affected by fluctuational factors $f_j(t)$. The three processes are represented schematically in Fig. 7.1: the observed process $y(t)$ by the thin line, the model process $z(t)$ – the basis of the forecast – by the heavy line, and the real process $x(t)$ by dots.

It is logical to take value y^0 of observation $y(t)$ at the starting instant t^0 as the initial condition z^0 for the model process $z(t)$:

$$z^0 \equiv y^0, \quad z^0 \equiv z(t^0), \quad y^0 \equiv y(t^0). \qquad (7.4)$$

It is due to this relation that the instrument noise $v(t)$ considerably affects prediction. Indeed, it can be deduced from (7.4) and (7.1) that the initial value x^0 of the model process is distinct from the initial value z^0 of the real process, $z^0 = x^0 + v^0$, and that, in contrast to equation (7.2) for $x(t)$ featuring local instability, like most cases of dynamical systems, the initial condition $v^0 = z^0 - x^0$ gives rise to an exponentially growing difference $|z - x| \approx v^0 \exp(\lambda_+ \tau)$, where λ_+ is the highest Lyapunov exponent and $\tau = t - t^0$ is the observation time.

The next cause of prediction errors is the impact of fluctuational forces $f_j(t)$ which produce a slight initial divergence between processes

$z(t)$ and $x(t)$ that later grows exponentially. This mechanism of error introduction is inherent in all physical systems without exception.

The third and perhaps the most important source of prediction errors is the inexactness of the model operator M_z with respect to the real operator M_x. It is an adequately chosen M_z that mostly determines success in prediction. What should we rely on in picking M_z? What factors limit the length of time for which behavior is predictable? Why do forecasts seldom come true? To answer these questions, we need to introduce a quantitative measure of prediction merit and discuss some notions related to the predictability of dynamical processes, that is, the degree of predictability, the time for which behavior is predictable, the predictability horizon, and so on.

7.3 Degree of Predictability. The Predictability Horizon

A universally adopted prediction accuracy characteristic is the mean-square error $\eta = y - z$:

$$\langle \eta^2 \rangle = \langle (y - z)^2 \rangle. \tag{7.5}$$

Here angle brackets denote empirical averaging. Assuming that $y_j = y(t_j^0 + \tau)$ is the value of $y(t)$ within period τ of the instant t_j^0 of the j-th observation, and that $z_j = z(t_j^0 + \tau)$ is the corresponding forecast for the period $(t_j^0, t_j^0 + \tau)$ satisfying the initial condition (7.4), $z_j^0 = y_j^0$, the mean-square error (7.5) is written as

$$\langle \eta^2 \rangle = \frac{1}{N} \sum_{j=1}^{N} \left[y(t_j^0 + \tau) - z(t_j^0 + \tau) \right]^2. \tag{7.6}$$

It is supposed that the greater the number of observations performed the more faithful is the error estimate (7.6). To specify potential predictability, it is convenient to use a dimensionless characteristic

$$D(\tau) = \frac{\langle yz \rangle}{\sqrt{\langle y^2 \rangle \langle z^2 \rangle}}, \tag{7.7}$$

which is simply a coefficient of correlation between the observed process and the model one at the time moment τ after observation has started. It equals unity at $\tau = t - t^0 = 0$, for the initial value of prediction z^0 is

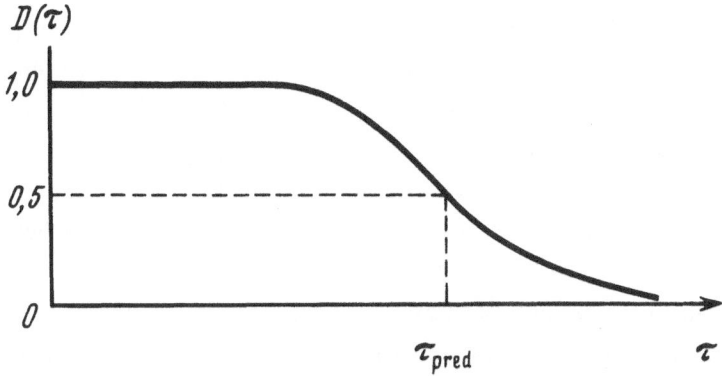

Fig. 7.2. Degree of predictability D plotted as a function of observation time $\tau = t - t^0$

taken to be equal to y^0. Reckoning y and z values from the mean level, we can see that D reaches zero with time.

The general pattern of the $D(\tau)$ relationship is depicted in Fig. 7.2. The values of $D(\tau)$ that are close to unity relate to a satisfactory forecast, while small values are the result of a discrepancy between observation and prediction. We shall therefore call $D(\tau)$ the *degree of predictability* and the time τ_{pred} in which $D(\tau)$ falls to 0.5 the *time of predictable behavior*. The degree of predictability D is unambiguously related to the prediction's absolute error $\langle \eta^2 \rangle$:

$$D(\tau) = \frac{\langle y^2 \rangle + \langle z^2 \rangle}{2\sqrt{\langle y^2 \rangle \langle z^2 \rangle}} \left[1 - \frac{\langle \eta^2 \rangle}{\langle y^2 \rangle + \langle z^2 \rangle} \right]. \tag{7.8}$$

Time τ_{pred} conforms to the absolute error

$$\langle \eta^2 \rangle \approx \frac{\langle y^2 \rangle + \langle z^2 \rangle}{2} \approx \langle y^2 \rangle,$$

that is, of the same order as the observed process variance.

In general, the degree of predictability $D(\tau)$ can be regarded as the degree of similarity between the true process and what we believe it is. Assuming that a high degree of similarity $D \approx 1$ represents determinate behavior, while small $D(\tau)$ values characterize unpredictable random behavior, $D(\tau)$ can be treated as the degree of determinateness of the observed process $y(t)$ with respect to the model $z(t)$. This interpretation was used in the author's earlier work [7.1]. The idea of identifying randomness and non-predictability underlies the conception of partially determinate processes and fields that was propounded in [7.1–3]. According to this conception, $y(t)$ is regarded as a fully determinate (predictable) process when the time interval τ concerned is small in comparison with the time of predictable behavior τ_{pred} (the predictability range), $\tau \ll \tau_{pred}$; as a fully random process when τ is large, $\tau \gg \tau_{pred}$; and as a partially determinate process in the intermediate case when $\tau \approx \tau_{pred}$.

The interpretation of randomness as non-predictability is just one of many ways to treat it in the language of modern science. In the set-theoretic approach upon which the modern theory of probability is based, randomness is associated with the existence of a probabilistic measure (a probability distribution), and all quantities, processes, and fields are classed as random whenever they possess a probability measure. Kolmogorov's algorithmic approach implies that randomness is identical to algorithmic complexity. In experimental physics, randomness is usually treated as a decaying correlation. For instance, let the process $y(t)$ be described by the correlation coefficient

$$K_y(\tau) = \frac{\langle y(t)\, y(t-\tau) \rangle}{\langle y^2 \rangle}. \tag{7.9}$$

Then states separated by a time interval exceeding the correlation time τ_c are regarded as random ones.

It is important to note that one and the same process may be random from the standpoint of one assumption and determinate from that of another. An example is a process whose complicated algorithm is random in terms of the algorithmic approach yet determinate from the standpoint of the theory of partial determinateness, for with its known algorithm at hand we can forecast $z(t)$ for sufficiently long times with $D=1$ guaranteed. Another example is a process for which the predictability range τ_{pred} exceeds the correlation time τ_c, where $y(t)$ can be regarded as random if randomness is assumed to be identical to the loss of correlation but determinate from the viewpoint of the partial deter-minateness theory identifying randomness with non-predictability.

The relationship $\tau_c < \tau_{pred}$ is fairly typical of many dynamical systems (examples will be given below). Moreover, as far as systems that are dynamically predictable at least in principle are concerned, the correlation time τ_c can be claimed to be *the least* (the worst) predictability range, and the correlation coefficient $K_y(\tau)$ to be *the least* (the worst) degree of predictability. The lower limit $\tau_{pred} = \tau_c$ is realized when no dynamic equation is available and the prediction is based on the principle that 'tomorrow will be like today'. In this case the forecast $z(t) = z(t^0 - \tau)$ is made using the value $y(t^0)$ observed at a given instant t^0,

$$z(t - \tau) = y(t^0) \quad \text{or} \quad z(t) = y(t - \tau), \tag{7.10}$$

and the degree of predictability (7.7) turns into the correlation coefficient (7.9). Apparently, this also applies to the coherence degree and time, since the correlation coefficient (7.7) does double duty as a measure of coherence.

Incidentally, all the attempts ever made to obtain a value of τ_{pred} markedly exceeding τ_c by statistical forecasting methods based on autoregression-type linear algorithms have been in vain. The comparability of τ_{pred} and τ_c can only be presumed from general considerations.

By way of example, consider linear forecasting on the basis of the $(m+1)$-dimensional representation

$$z(t^0 + \tau) = a_0 y(t^0) + a_1 y(t^0 - \tau) + \ldots + a_m y(t^0 - m\tau) \tag{7.11}$$

composed of $m+1$ prior observation values. With optimal coefficients a_m chosen such that the mean-square prediction error (7.5) is minimal, all a_m are dependent on the correlation function $K_y(n\tau)$, $n = 0, 1, \ldots, m$. At the same time, the degree of predictability (7.7) is expressed in terms of the same $K_y(n\tau)$ values. As a result, the equation $D(\tau) = 0.5$ from which τ_{pred} is to be determined contains merely the correlation function values and fails to provide a value of τ_{pred} differing significantly from τ_c. As far as I can judge, these simple considerations have not yet been fully accepted by those who devise linear statistical predictors of the autoregression type, although convincing evidence of their validity exists in the literature.

One of the cases in point is noisy dynamical chaos [7.1, 7.4]. Suppose that λ_+ is the highest positive Lyapunov exponent. Then the mean-square prediction error grows exponentially:

$$\langle \eta^2 \rangle = \left(\sigma_v^2 + \sigma_f^2 + \sigma_{\Delta M}^2 \right) \exp(2\lambda_+ \tau), \tag{7.12}$$

with σ_v^2, σ_f^2, and $\sigma_{\Delta M}^2$ representing the contributions of instrument noise, internal physical noise, and the impact of model inaccuracy $\Delta M =$

$M_x - M_z$, respectively. The summand $\sigma^2_{\Delta M}$ is sometimes called the contribution of 'ignorance noise' (the term used in the theory of weak signal detection). Equating the mean-square error (7.12) with the observation variance $\langle y^2 \rangle$, we can estimate the time of predictable behavior:

$$\tau_{pred} \approx \frac{1}{2\lambda_+} \ln \frac{\langle y^2 \rangle}{\sigma^2_v + \sigma^2_f + \sigma^2_{\Delta M}}. \qquad (7.13)$$

Quantity $1/2\lambda_+$ is comparable with the system's correlation time in terms of its order of magnitude: $\tau_c \approx 1/2\lambda_+$. Therefore the ratio τ_{pred}/τ_c is characterized by the value of the quantity whose logarithm appears in (7.13). For example, if the 'noise' summands σ^2_v, σ^2_f, and $\sigma^2_{\Delta M}$ equal one-thousandth of the mean-square amplitude

$$A_y \approx \langle y^2 \rangle^{1/2}.$$

Then

$$\tau_{pred} \approx \tau_c \ln 10^6 \approx 14\tau_c,$$

that is, the predictability range is fourteen times as long as the correlation time.

The essential notion of the predictability horizon, τ_{hor}, was introduced by James Lighthill [7.5] to denote the time interval in which a microscopically small discrepancy in the initial conditions grows to a macroscopic value of the order of A_y. I suggest treating the predictability horizon in a somewhat different way, as a finite timespan of predictable behavior that cannot be surpassed by either improved measuring instruments or a refined prediction model $z(t)$ [7.1]. In other words, τ_{hor} should correspond to the limit of negligibly small measurement and ignorance noise:

$$\tau_{hor} = \lim_{v \to 0, \Delta M \to 0} \tau_{pred}. \qquad (7.14)$$

In the case of noisy dynamical chaos we have just discussed, this condition is met by regarding the contributions of σ^2_v and $\sigma^2_{\Delta M}$ as negligibly small in comparison with that of physical noise σ^2_f, which can never be completely eliminated:

$$\tau_{hor} = \frac{1}{2\lambda_+} \ln \frac{\langle y^2 \rangle}{\sigma_f^2}.$$ (7.15)

Based on the above reasoning, we can generally look into the ways in which the degree of predictability changes with improving prediction model $z(t)$. For short times not exceeding the correlation time τ_c, good forecasts are made using the principle that 'tomorrow will be like today' [7.1]. In these events, the degree of determinateness (7.7) coincides with the correlation coefficient (7.9), whose time variation is shown by curve 1 in Fig. 7.3. At $\tau_{pred} \approx \tau_c$, some improvement in forecasting quality is provided by linear statistical models of the autoregression type, but a real breakthrough is possible only with model processes derived from dynamical equations.

As the model is refined, the degree of predictability grows, along with the time of predictable behavior (Fig. 7.3). At this stage, a random unpredictable process $y(t)$ (point A on curve 2 referring to model $z_1(t)$) may turn into a determinate predictable one (point B on curve 3 referring to a refined model $z_2(t)$). Such a transition is possible only in terms of a concept of partially determinate processes that identifies randomness with non-predictability and is used in physical experimentation. As a

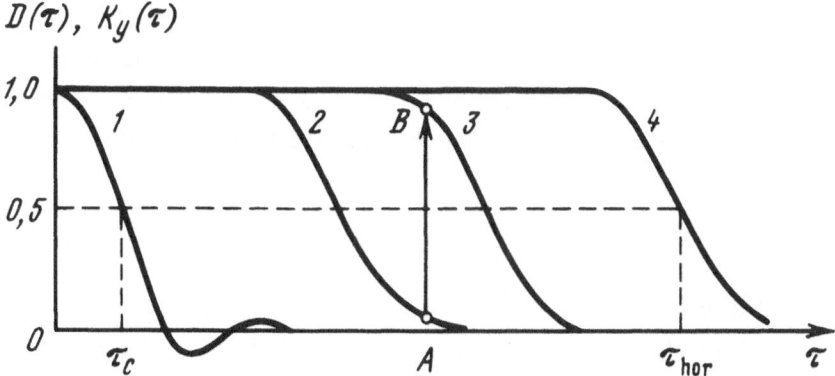

Fig. 7.3. Degree of predictabity changing with model improvement. Curve 1 presents the quality of prediction based on the principle that 'tomorrow will be the same as today', curve 2 presents the quality of prediction for a model $z_1(t)$, curve 3 presents the same for an improved model $z_2(t)$, and curve 4 presents the largest achievable time of predictable behavior (the predictability horizon). The $A \rightarrow B$ transition from $z_1(t)$ to $z_2(t)$ carries the process over from the category of random (unpredictable) events to that of determinate (predictable) ones

matter of fact, we are speaking here of something initially un-comprehended that then becomes understood. In the language of an axiomatic theory of probabilities, such a transition is impossible in principle, for there a random process remains random irrespective of whether there is a model whereby it can be predicted.

Finally, there is a limiting prediction time, that is, a predictability horizon τ_{hor}, beyond which no model can provide an adequate forecast. A limiting degree of predictability is illustrated by curve 4 in Fig. 7.3.

Estimating a predictability range is a matter of principal importance, since attempts to make predictions for times exceeding τ_{hor} are totally senseless. They bring only losses in time and funds, to say nothing of disappointment. In those cases where for all tested model processes $z(t)$ it is impossible to achieve a high degree of predictability $D(\tau)$ for times exceeding the domain of the correlation coefficient $K_y(\tau)$, i.e., where $\tau_{pred} \leq \tau_c$, it should be admitted that a dynamic model for process $y(t)$ is not available and that $y(t)$ should be categorized as noise. In other words, noise is a stochastic process (in the general sense of the theory of probability) to which no dynamical prediction model can be fitted. At the same time, processes that are predictable for times exceeding the correlation time are to be classed with partially determinate ones.

In contrast to the above evolution algorithms described by differential equations, there are finite-difference (mapping) algorithms operating in many systems. Let us summarize what we know about the latter, for cases to be discussed below have to do precisely with finite differences.

Suppose that x_m is a dynamical variable, $F(x_{m-1}, f_{m-1})$ is a nonlinear function converting x_{m-1} into x_m, and f_{m-1} are fluctuational effects. Then a real sequence is governed by

$$x_m = F(x_{m-1}, f_{m-1}). \tag{7.16}$$

The observed sequence $\{y_m | m = 0,1,...\}$ differs from $\{x_m\}$ due to nonlinear, spectral, and noise-induced distortions. In the simplest case, where only additive noise is to be considered, y_m is related to x_m by an expression $y_m = x_m + v_m$ similar to (7.1).

Let us use a 'noise-free' map

$$z_m = F_z(z_{m-1}, 0) \tag{7.17}$$

in place of the differential model equation (7.3). The starting value of the observed process, y^0 will be used as the initial condition z^0 of the model sequence: $z^0 = y^0$.

At this point, in order to validate the general statements appearing in this section, let us introduce some formulas suggesting the exceptional importance of noise and the no less important role of modeling in what concerns prediction of locally unstable systems. Figures 7.5–7 show computed degrees of predictability D and correlation coefficients K_y as reported in [7.2]. They were obtained for a system described by a one-dimensional linear piecewise mapping of segment (0, 1) into itself:

$$x_m = F(x_{m-1}) + f_{m-1}, \quad F(x) = \{Ax\}, \quad y_m = x_m. \tag{7.18}$$

Here $F(x) = \{Ax\}$ is a fraction of Ax (diagrams for $z_m = F(z_{m-1})$ at $A = 1.1$, $A = 2.0$, and $A = 9.9$ are shown in Fig. 7.4, parts a, b, and c, respectively), and f_m is an additive fluctuational effect whose probability density is uniform in the range $-10^{-p} \le f \le +10^{-p}$, where p is a fluctuation intensity exponent such that the greater the value of p the lesser the fluctuations. In our examples, $2 \le p \le 6$. Values of f_m and f_n, for $m \ne n$, are considered to be uncorrelated: $\langle f_m f_n \rangle = 0$. In using a model of additive noise f_n, it seems unnecessary to allow for measurement noise, and no distinction is made in (7.18) between the observed and real sequences: $y_m = x_m$. As for the model sequence z_m, it obeys a noise-free equation

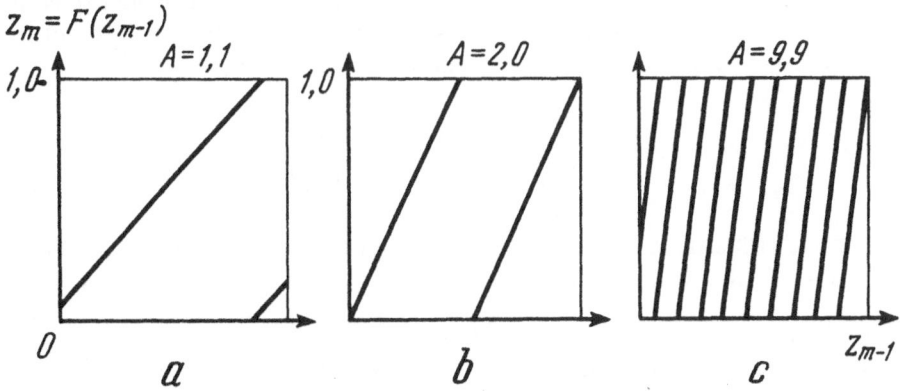

Fig. 7.4a–c. A linear piecewise mapping: (a) for a low slope (A = 1.1), (b) for a moderate slope (A = 2.0), (c) for a steep slope (A = 9.9)

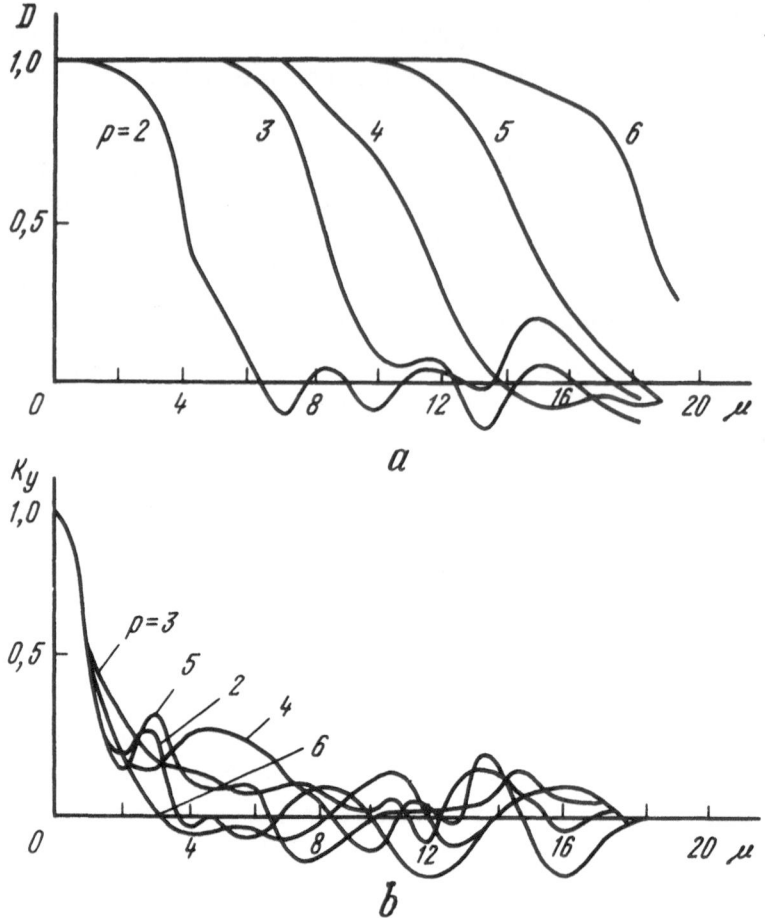

Fig. 7.5a,b. Degree of predictability D and coefficient of autocorrelation K_y as functions of discrete time μ, for different noise strengths $\sigma_f = 10^{-p}/\sqrt{3}$ (the greater the value of p the lower the noise strength). It is important that the degree of predictability is strongly dependent on noise strength, whereas the sensitivity of the autocorrelation coefficient $K_y(\mu)$ is insignificant

$$z_m = F(z_{m-1}), \quad F(z) = \{Az\}. \tag{7.19}$$

Figure 7.5 illustrates a numerical modeling of the degree of predictability $D(\mu)$ as a function of the discrete time $\mu = m - m^0$ elapsed since the beginning of the tests. The sequence of curves in Fig. 7.5a shows five cases where the slope coefficient $A = 2$ (see Fig. 7.4b) and $p = 2,\dots,6$, respectively, i.e., five noise levels from $\sigma_f = 10^{-2}/\sqrt{3}$ to $\sigma_f = 10^{-6}/\sqrt{3}$.

As indicated by the curves in Fig. 7.5a, the degree of predictability is close to unity for 5–15 time steps, whereupon it quite quickly drops to zero. The duration of predictable behavior is strongly dependent on the noise level σ_f and obeys a logarithmic relationship of type (7.15) with Lyapunov exponents $\lambda_+ = A$. At $A = 2$, $\mu_{pred} \approx (1/2A)\ln(1/\sigma_f^2)$ in order of magnitude, which in fact agrees well with actual values of μ_{pred} taken from Fig. 7.5a.

Meanwhile, the correlation coefficient $K_y(\mu)$ goes down to around 0.5 at practically the first step and displays a fairly weak sensitivity to noise (Fig. 7.5b). This result demonstrates the profound difference between correlation and predictability times.

The changing parameter A in mapping (7.17) considerably affects the quality of the prediction. Figure 7.6 (a and c) illustrates $D(\mu)$ functions for a map with small ($A = 1.1$), medium ($A = 2.0$), and large ($A = 9.9$) slope and two noise strengths: Fig. 7.6a shows the case of a moderate noise level ($p = 2$) where noise accounts for around 1% of a typical

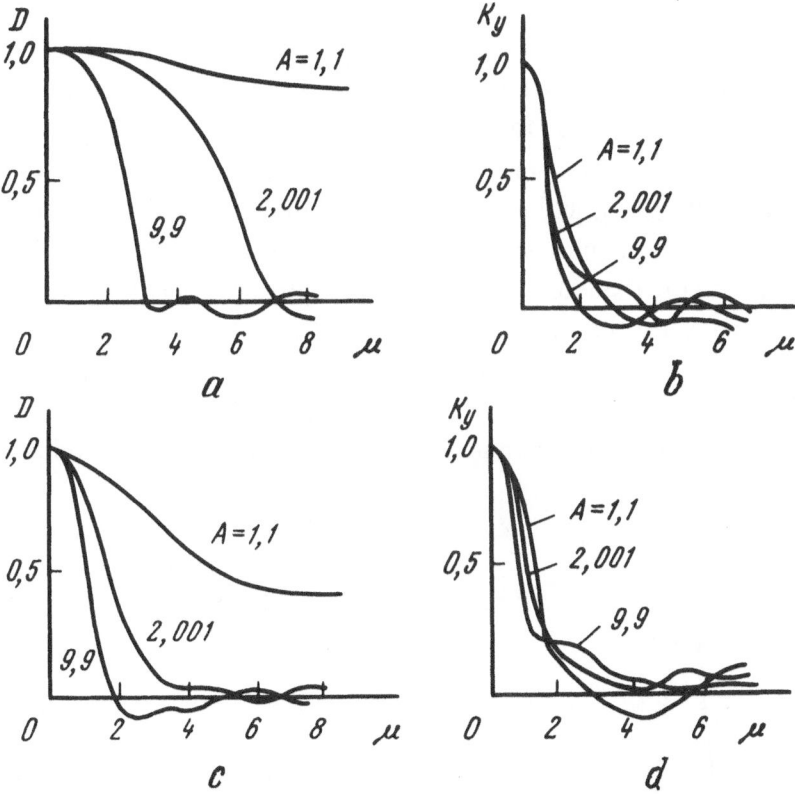

Fig. 7.6. (a, b) Degree of predictability D and coefficient of autocorrelation K_y as functions of discrete time μ, for a moderate noise strength ($p = 2$, $\sigma_f = 0.01/\sqrt{3}$) and for different slopes $A = 1.1, 2.0, 9.9$. (c, d) The same, but for a stronger noise ($p = 1$, $\sigma_f = 0.1/\sqrt{3}$)

$y \approx 1$, while Fig. 7.6c shows a case of strong noise ($p = 1$) that accounts for 10% of the observed oscillations. The figures show how predictability time diminishes with the map slope A. For example, as Fig. 7.6c indicates, with strong noise ($p = 1$) the predictability range diminishes from 4.7 at $A = 1.1$ to 1.4 at $A = 9.9$. Meanwhile, parts b and d of Fig. 7.6 depicting the behavior of $K_y(\mu)$ suggest that in going from a small slope ($A = 1.1$) to a large one ($A = 9.9$) and from low noise ($\sigma_f \approx 10^{-2}$) to high noise ($\sigma_f \approx 10^{-1}$) the correlation time remains at the practically unchanged value of about one step.

A slight variance ΔA of the coefficient A appearing in the model equation (7.19) affects the prediction range at $\Delta A \geq \sigma_f$, since, according to (7.13),

$$\mu_{pred} \approx \frac{1}{2A} \ln \frac{1}{\sigma_f^2 + (\Delta A)^2} \, .$$

Investigating the dependence of μ_{pred} on the hypothetical value $\tilde{A} = A + \Delta A$ (Fig. 7.7), or – more practically – the dependence of the predictability degree D on $\tilde{A}$ for fixed μ, accomplishes two useful tasks: (1) estimation of the parameter A in a physical system from, say, a value enabling maximal μ_{pred}, and (2) estimation of the noise level in system σ_f from the width of the curve in Fig. 7.7. Both tasks are very often involved in practical studies, and not only in physics. It is a matter of principal importance that amplification of noise σ_f and inaccuracy of model parameters ΔA are given equal consideration, for it may now be argued that if no changes can be made in a real physical system, this kind of uncertainty cannot be eliminated and no refinement of the model is helpful.

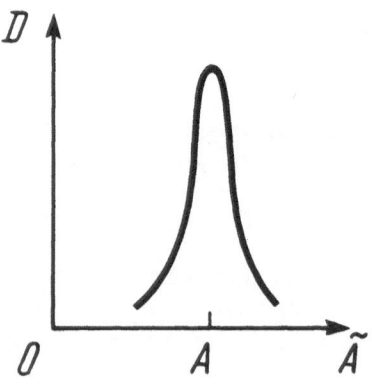

Fig. 7.7. Degree of predictability D as a function of the slope $\tilde{A}$. The width of the curve is governed by the noise strength

7.4 Searching for Prediction Models

From the standpoint of mathematics, the predictability problem reduces to solving the inverse problem, that is, reconstructing the dynamical equation governing an observation $y(t)$ [7.7]. With such an equation at hand, solutions are possible for $t > t^0$. In other words, predictions $z(t)$ can be obtained.

Solving the inverse problem presents immense difficulties. To begin with, there is an infinite number of equations which observation $y(t)$ may obey to within the accuracy allowed by the fluctuational conditions of the problem. This is how the incompleteness of any system of hypotheses, models, or idealizations manifests itself in accordance with Gödel's theorem on the incompleteness of any system of axioms.

Clearly, the inverse problem is unsolvable if we do not somehow constrict the set of candidate equations. It is actually this constriction that presents the basic difficulty. However, if the structure of the system involved is known in outline then the problem reduces to determining the coefficients of these equations. If not, it is reasonable to test a few candidate models to see if they explain the observed behavior. General requirements on models are discussed again and again in the literature. Eventually the discussions all boil down to this: a model that is neither too sophisticated nor oversimplified must adequately render the basic properties of the system. In effect, the construction of a model is an exercise in natural science. There is, perhaps, one unusual point about these particular inverse problems: in attacking them, we aim to elucidate a system's dynamic structure in terms of familiar regularities rather than by searching for fundamental underlying mechanisms.

Apart from the individual treatment of each specific case by means of a profound investigation of the physical processes taking place in the system (the only option available until recently), one can use a more formal mathematical approach involving the use of a supercomputer. What I mean is setting up a library of typical dynamical equations whose coefficients can be automatically adjusted for the observed process. This procedure, however time-consuming it might be, is promising.

When choosing typical dynamical equations, one should bear strongly in mind those systems already investigated that are apt to display chaotic behavior (strictly periodic systems are of little practical interest). There are no less than a hundred such systems that have been investigated so far. Among them are the Lorenz attractor, Duffing's driven equation, and nonlinear oscillatory motions caused by a strong periodic force.

Furthermore, each of the typical equations contains several (say, $n = 3,4,5$) parameters whose variations allow adjustment to suit the observed process. Let us assume that Δa_k is a reasonable increment for the variation of a_k and $A_k = \max a_k - \min a_k$ is the range of variation of parameter a_k among the equations. The exhaustive search algorithm involves calculating the estimate

$$S = \prod_{k=1}^{n} \frac{A_k}{\Delta a_k}$$

of the number of solutions to these typical equations.

The resulting number of solutions that are to be tested for their closeness to the observed process may be extremely large. For example, if $A_k/\Delta a_k = 100$ (an underestimate rather than an overestimate), while the number of varying parameters is five, then $S \cong (100)^5 = 10^{10}$. This result should now be multiplied by the number Q of tested types of equation. Given $Q = 100$, say, the total number of solution variants $S_1 = QS \cong 10^{12}$.

It seems impossible to exhaust such a number of variants even with a powerful computer. Therefore, we should do our best to reduce it. There are two more or less obvious opportunities for doing so.

– *A preliminary choice of a constricted class of equations* based on expert estimates. Very likely, the number of equations under test can be reduced three- to tenfold in this manner. Of course, this is not very much, considering the value of 10^{12}, but not too bad either for a start.

– *Employment of adaptive and self-complicating algorithms.* Instead of an exhaustive search for all possible relevant parameters, one can try to select a crude model operating over comparatively short times and then to specify a smaller number of parameters in a relatively narrow range of values. As a matter of fact, we are attempting to employ algorithms of the type used to create a chess-playing program. The most important features of these algorithms, the discarding of *a priori* senseless variants, is of great interest for the predictability problem.

Comparing the problem of dynamical systems prediction with that of chess algorithms, there is one point that makes the former look simpler – the absence of an active competitor imposing his/her/its own behavior pattern. However, an active countervailing force is also to be found for prediction models, namely noise, which introduces considerable uncertainty.

With regard to the general problem of reconstructing a system's dynamical structure, let me summarize some techniques that are helpful in building dynamical models.

1. It is reasonable for practical purposes to employ averages over the observation period T rather than to consider local mean-square errors $\langle \eta^2 \rangle$ and predictability degrees $D(\tau)$. Denoting the combined averages (over realizations and time) by braces, for example, as

$$\{yz\} = \frac{1}{T} \int_{t^0}^{t^0+T} \langle y(t')z(t') \rangle \, dt' ,$$

we can speak in terms of quantities

$$\{\eta^2\} = \{|y - z|^2\} \quad \text{and} \quad D(\tau) = \frac{\{yz\}}{\{y^2\}^{\frac{1}{2}}\{z^2\}^{\frac{1}{2}}} , \tag{7.20}$$

which serve as a sort of filter of the observed process $y(t)$ relative to the model process $z(t)$. As a result, it even becomes possible to detect a weak signal $x(t)$ obscured by noise.

2. The model equation (7.3) can be used as a figure of merit of adjustment, provided observation $y(t)$ enters into (7.3). It implies estimation of quantity

$$s_1 = \left| M_z\left[\frac{d}{dt}, \, y(t)\right] \right| . \tag{7.21}$$

Squaring s_1 averaged over time T,

$$s_2 = \frac{1}{T} \int_{t^0}^{t^0+T} \left[M_z\left[\frac{d}{dt}, y(t)\right] \right]^2 dt , \tag{7.22}$$

is also appropriate. Quantity s_2 operates as a kind of model filter; in those cases where observation $y(t)$ does not properly fit the model equation $M_z(d/dt, z) = 0$, s_2 shows up as unacceptably large and is therefore discarded.

Sometimes it is more convenient to minimize $\{\eta^2\}$ at the initial stage of rough adjustment and to search for minimal values of s_1 or s_2, which are especially sensitive to changing parameters, at later stages.

3. Some techniques from so-called discriminant analysis can be recommended. This kind of analysis is helpful in detecting internal alterations to the process which are hard to reveal by estimation or by using elementary criteria of the variance-constancy type. These techniques have to do with the analysis of nonlinear combinations containing observation $y(t)$, its orders $y^2, y^3, \ldots$, and its derivatives $y', y'', \ldots$ The most informative combinations are those that are approximately constant (the approximate invariants) over some time interval, say

$$y'' + a_1 y' + a_2 (y')^2 + a_3 yy' + a_4 y^2 + a_5 y^3 \cong \text{const.} \qquad (7.23)$$

Any alterations in the value of (7.23) that may occur should point to variations of the parameters a_k, that is, to changes taking place within the system. The search for invariants, as applied to the problem under discussion, actually reduces to the determination of an approximate differential equation for a given system, which makes such a search worthy of special consideration.

4. It must be emphasized with reference to the inverse problem case that the role of noise is contradictory. Paradoxical as it might seem, the higher the noise level, the simpler the design of a prediction model. The fact is that many equation summands lose significance with the growth of fluctuational forces and can be neglected, as a result of which the computer time needed for an exhaustive search of parameters is dramatically shortened (by orders of magnitude). Naturally, at high noise levels the duration of predictable behavior is small.

In this context, it is recommendable to introduce a low noise into the process of selecting suitable differential equation coefficients. Such noise may regulate the process of step-by-step approximation.

An important point is that introducing artificial noise also enables one to estimate a system's internal fluctuations. To accomplish this, one has to fix the noise level at which the coefficients of a reconstructed differential equation start to display relative independence.

Naturally, much depends on the spectral density of the artificial noise (its frequency spectrum 'color') and on exactly where (in which summand) it is embedded. Nevertheless, the mere possibility of estimating the level of the fluctuational forces, however rough it may be, is appealing.

5. There are prediction techniques that involve no reconstruction of differential equations, for example, prediction using patterns from the past that are similar to those of today. Suppose that a system's behavior before a specific instant t^0 looks very much like that before instant $t*$ in the past. Then the required forecast for $t > t^0$ is $z(t^0 + \tau) = y(t* + \tau)$, and there is no need to model the system's dynamical structure. In this way, the data on historical behavior is in effect used as an analogue solution. This is exactly the way that long-term weather forecasting (six months ahead and longer) is done using weather patterns from the past.

This forecasting technique was tested quite recently by Farmer and Sidorovich for several chaotic system mappings [7.6]. The most important finding of these authors was that the actual time of predictable behavior was considerably greater (by factors of 3, 10, and even 100) than the correlation time τ_c. In my opinion, this very promising result requires further analysis and contemplation. Specifically, it is not quite clear what exactly influences the prediction range, noise $f(t)$ within the system, measurement noise $v(t)$, or inaccuracy of initial conditions at instants t^0 and $t*$. Besides, it remains to be explained how the general instability (variability) of external conditions affects predictability.

Naturally, this instability issue pertains to all methods of prediction without exception, and we shall discuss it briefly in the next section. But at this point let us emphasize the necessity of elaborating figures of merit for a stationary system to be forecasted using a stationary (time-independent) model. Such figures of merit are required to assess the validity of algorithms picked for a 'trial' time interval as well as their applicability to times beyond this interval.

6. It is worthwhile to outline the advantages and disadvantages of the 'forecasting into the past' method sometimes recommended to check prediction algorithms obtained using a limited database. The method involves comparing observation $y(t^0 - \tau)$ and prediction $z(t^0 - \tau)$ for the instant $t^0 - \tau$ *prior* to the observation instant t^0. It stands to reason that forecasting into the past is possible only if the algorithm admits a one-to-one correspondence between the past and the future. Such a condition immediately rules out mappings lacking this correspondence

for cases of finite difference equations, as well as hysteretical systems for cases of strongly nonlinear differential equations.

However, even with the one-to-one correspondence, the 'backward' prediction range does not necessarily coincide with the 'forward' prediction range, since fluctuation build-up is not the same in the forward and backward directions. This is evident from expressions similar to equation (7.13) that provide estimates of the prediction range for a locally unstable system with the highest possible Lyapunov exponent λ_+. In using this expression for predicting into the past, λ_+ is substituted by the least negative exponent λ_- to characterize the inverted-time fluctuation build-up. Incidentally, the finite duration of predictable behavior is a fact of life for historians, archeologists, detectives, and many other specialists in 'backward' prediction. Among the factors limiting predictability, an important one is the nonstationary character of dynamical equations of motion mentioned above.

7. In working some problems it is possible to go over from an algorithm in the form of differential equations to a finite difference algorithm. This is accomplished using a Poincaré mapping which fixes the relationship between successive points ξ_m at which a phase trajectory intersects the Poincaré plane. Clearly, this sort of mapping reduces dimensionality; the dimension of vector ξ on the Poincaré secant plane is less by one than that of the total phase space.

Once the initial problem is reduced to a one-dimensional mapping, we can actually consider the problem to be solved, for the properties of a noisy one-dimensional mapping are fairly well understood. By way of example, consider one of the earliest problems where dynamical structure is restored at a mapping level [7.8].

The system concerned is a self-oscillator with a tunnel diode (a Kiyashko–Pikovskii–Rabinovich oscillator) generating trains of exponentially growing sinusoids which stop growing at the instant when the diode becomes open. The generated trains of pulses are represented schematically in Fig. 7.8. The entire problem is described by a third-order differential equation. However, in some circumstances, it reduces to the study of a mapping which relates sequential local maxima y_m between themselves. Typical configurations are made by experimental points in the plane (y_m, y_{m-1}) (Fig. 7.9). They are located near certain center lines $y_m = \overline{F}(y_{m-1})$ that can be taken as a basis for the forecast $z_m = \overline{F}(z_{m-1})$.

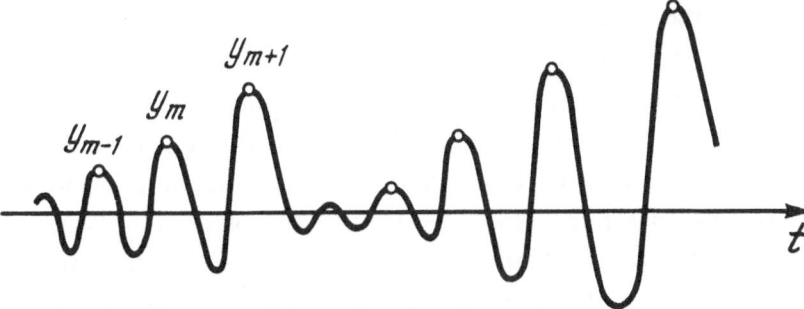

Fig. 7.8. A typical pulse train diagram for the Kiyashko–Pikovskii–Rabinovich (KPR) oscillator

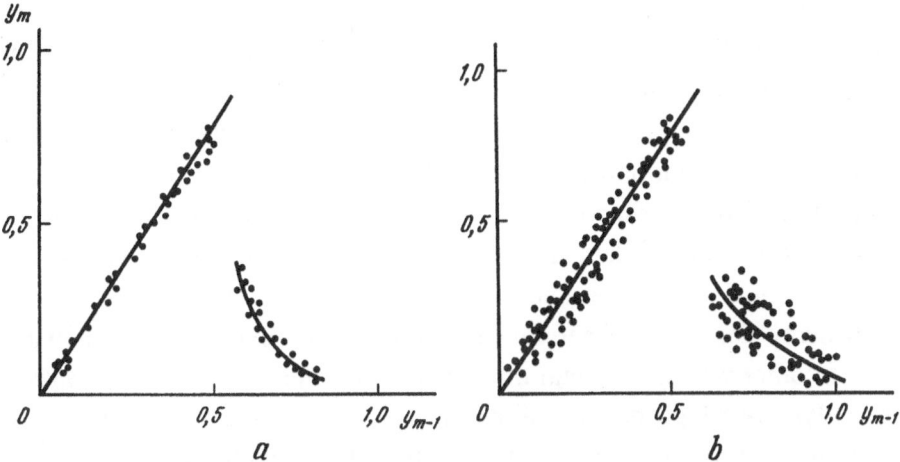

Fig. 7.9. Mapping in the case of the KPR oscillator: dots denote experimental values and solid lines show model predictions

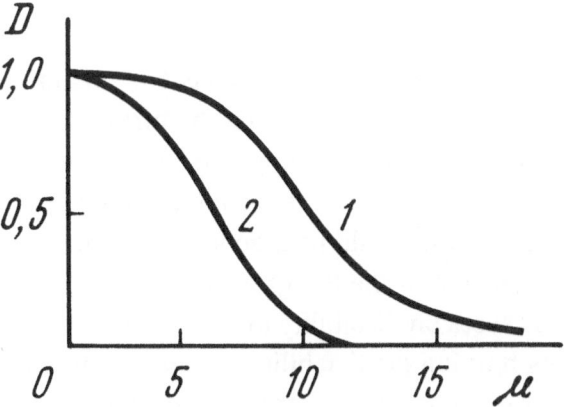

Fig. 7.10. The degree of predictability for the KPR oscillator: curve 1 refers to natural noise and curve 2 to enhanced noise

Practically, the curve was plotted using polynomial approximation (individually for each of the two portions), with the polynomial coefficients determined by the least-squares method. The degree of determinateness of the forecast z_m based on the curve $\overline{F}(z_{m-1})$ is shown in Fig. 7.10 by curve 1, which refers to the case of natural noise occurring within a self-oscillator (see Fig. 7.9a) when the predictability range equals about 10 cycles. With additional noise introduced into the oscillator, the scatter of experimental points on the plane (y_m, y_{m-1}) increases (Fig. 7.9b), though the model curve $y_m = \overline{F}(y_{m-1})$ remains practically undeformed. As might have been expected, the additional noise shortens the predictability range almost twofold, from 10 to 5 cycles (curve 2 in Fig. 7.10).

In concluding this section, it must be emphasized that information about the dynamical chaos fractal dimension may be helpful for constructing prediction models. Among other things, a fractal dimension can define the approximate number of degrees of freedom involved in a motion, and hence the order of the differential equation providing a model. Nonetheless, data derived from a fractal dimension should be treated critically. For example, one can find a pretty complicated system with many degrees of freedom behind a limiting cycle with dimension one. In other words, a fractal dimension characterizes topological rather than dynamical properties of a system. This fact should lead us to examine carefully the entire database available to us before we consider the fractal dimension of a model system.

7.5 Limits to Predictability

In the previous section we concentrated upon methods for obtaining dynamical system prediction models and summarized the difficulties to be overcome, primarily through the use of computers. But however helpful computers might be, the difficulties of prediction are not merely due to insufficient computer power. It would be a good idea to look for more critical limitations in the nature of the things we predict. So, what exactly in the nature of things causes predictability to deteriorate?

An enumeration of factors limiting predictability would require a lot of space here. Nevertheless, we cannot pass over those of fundamental significance. They include the local instability of many of the systems under review, distortion of predictions due to measurement effects, evolutionary processes in open systems, bifurcation points traversed in noisy systems, and forecasting costs. Let us discuss these more closely.

Local Instability. The feature of local instability complicates prediction, especially when perturbations grow exponentially, as happens with dynamical chaos. In this event the predictability range, according to equations (7.13) and (7.15), depends logarithmically upon dynamical and fluctuational parameters. This means, for example, that to increase τ_{pred} by an order of magnitude (i.e., tenfold) the signal-to-noise ratio SNR = $\langle y^2 \rangle / \sigma_f^2$ must increase by e^{10}, that is, almost 10 000 times! This example shows why the predictability horizon is pushed back so slowly with signal amplification or noise attenuation.

From a very general standpoint, the local instability of classical systems can be regarded as the cause of randomness in the physical world. Microscopically weak perturbations of, say, a thermal nature affecting locally unstable chaotic systems can reach macroscopic proportions in finite time. In other words, dynamical chaos works as a gigantic amplifier of microfluctuations. As it does so, the observer is only able to trace the behavior of the resulting macrofluctuations, if anything. In this context, detection of local instabilities in a system under study is of prime importance, since the predictability horizon of such a system is especially small.

This deduction is confirmed by the example of a Boltzmann gas, that is, an aggregation of small elastic spheres modeling real gas molecules. In conditions typical of the Earth's atmosphere, a molecule's mean free path length l exceeds its diameter a by $q = l/a = 10^4$ times. This figure shows how many times a particle's angular perturbation $\delta\theta_m$ increases on collision with another particle, that is, $\delta\theta_m = q\delta\theta_{m-1}$. If there are several collisions, the local perturbation reaches $\delta\theta_m \approx q^m \delta\theta_0$. It is readily apparent that at $q = 10^4$, $\delta\theta_m$ can reach any value after a finite (and quite moderate) number of collisions have occurred. For example, if $\delta\theta_m \approx 1$ rad corresponds to a complete loss of memory about the initial direction of the particle motion, then

$$m \approx \frac{\ln(\delta\theta_m/\delta\theta_0)}{\ln q} \approx \frac{|\ln \delta\theta_0|}{\ln q}.$$

However small the initial perturbation $\delta\theta_0$ might be, the number m of collisions is less than ten, or a few tens at the utmost. This suggests that the predictability range for a Boltzmann gas encompasses just a few collisions. If it is remembered that two successive collisions are spaced approximately 10^{-10} s apart, then the predictability horizon τ_{hor} is no more than 10^{-9} s. Incidentally, this estimate seems to have a direct bearing on the problem of the irreversibility of behavior of a real gas: the motion of all molecules is reversible only over a time interval $\Delta t \leq \tau_{hor}$ from the moment when the velocities of all the molecules is reversed, whereas for $\Delta t \geq \tau_{hor}$ the particles inevitably lose their memory of their velocities before reversal occurred, for example, by having it erased by a weak thermal electromagnetic field.

With regard to hydrodynamic fluxes, the predictability horizon is of course much larger than for a Boltzmann gas, but the difference is not very tangible if τ_{hor} is considered with respect to the correlation time τ_c, for in this event the logarithmic dependence of the signal-to-noise ratio still holds. It is this stipulation that can explain our slow progress in weather forecasting.

Prediction Distortion Due to Measurement Effects. The problem of predictability can be formulated not only in terms of classical physics but also in the language of quantum mechanics. There it appears that the wave function of a system under review should be subject to prediction.

On a superficial view, there is nothing fundamentally new in comparison with a corresponding classical system: in place of one predictand (a classical trajectory) there is another (a wave function) to be considered. However, there is a difference, although it is not as fundamental as believed earlier, and it has to do with variations of the initial function Ψ^0 that are critical to prediction.

As we know from the theory of quantum measurement, the reading given by a measuring instrument and the actual state of the system are based upon different wave functions. Therefore, a forecast based on instrument readings diverges from real behavior. This divergence of measured initial conditions from the actual conditions is due to the distorting influence of the instruments.

Up until quite recent times, it was believed that the instrument effect had to be taken into consideration only for quantum systems. But effects due to measurement or – to be more exact – instruments (meant in a wide sense here) also pertain to classical systems if these feature local instability [7.1]. The problem is that a small perturbation introduced by an instrument in a locally unstable system can be rapidly amplified, causing a drastic distortion in the observed process. In this event, the readings that are taken do not indicate the undisturbed process we intended to measure. The distorting effect of instruments on measurable classical processes has not so far been analyzed in detail.

This distorting effect may also show up when an attempt is made to trace small-scale air or liquid motions using tiny sensors. It is clear that increasing the number of sensors will sooner of later result in distortion of the flow even at the 'classical' level. It seems reasonable to look for an optimal degree of interaction between the instrument and the classical system involved, so that a suitable trade-off between the amount of information derived and its distortion can be reached. Such considerations concerning a 'trade-off' measurement apply in full to quantum systems as well.

Evolution of Open Systems. The distinguishing feature of big open systems is their ability to undergo evolutionary changes: instabilities arise within them on one or another scale and new structures spring up and disappear, that is, self-organizational metamorphoses take place. There are a number of rather intricate questions bearing on evolutionary self-organizing systems that are not easy to answer definitively. In particular, they concern the following.

Firstly, in cases of evolving systems, transformations of their model equations and/or alterations in their solution patterns have to be taken into account. The fundamental equations of motion naturally remain unaltered, but model representations of a system's behavior at different stages of its evolution may be different. The pattern of solutions of the governing equations may also be different. For instance, the Airy equation $x'' - tx = 0$ has bounded oscillating solutions for $t < 0$, while for $t > 0$ it has exponentially amplifying and attenuating solutions.

Secondly, there are difficulties connected with working out a way to measure the quality of predictions under evolution conditions. It is usually possible to obtain a more or less adequate measure of quality for stationary systems, namely, the mean-square error $\langle \eta^2 \rangle$ or the degree of predictability $D(\tau)$. They are justifiable if prediction and observation are repeatedly correlated during a lengthy time period T.

It is different with evolving systems, in whose cases universally adopted measures of quality apply only to relatively small time scales: the total observation time T made up of m subintervals $\Delta t = T/m$ must be short as compared to the interval Δt_{ev} typical of evolution. If $T \geq \Delta t_{ev}$, the experimenter is unable to optimize measures of quality 'on the go' and has to restrict himself to single comparisons of observation and prediction.

Thirdly, the most troublesome problem is that of detecting instabilities occurring in the course of evolution. Time τ_{inst} characteristic of unstable processes is usually shorter than time Δt_{ev} typical of evolutionary changes (if this were not so, i.e., if $\tau_{inst} \geq \Delta t_{ev}$, then instability development would be indistinguishable from a background evolutionary process). Immediately after its origination, the exponentially growing instability remains hidden behind noise. However, it can be detected some time later, after it exceeds the noise level. The rapidly growing instability reaches macroscopic proportions in a finite time of the order of several τ_{inst} (this time is smaller for the so-called explosive-type instabilities).

Under these circumstances, instability can often be detected only a posteriori, after the initial forecast disregarding exponential instability has become invalid.

Fourthly, as forecasting skill is improved, one can anticipate a hierarchical extension of model equations in open systems due to integration of ever newer subsystems and effects that were neglected at small space-time scales. Small- and large-scale variations of the nearest neighborhood are equally important for making good forecasts. Small-scale events may show up as short-term noise and large-scale events are often responsible for a slow change of the parameters of one of the subsystems involved.

The hierarchical character of prediction models represents an important feature of the physical world, where *all things are interrelated*. The principle of universal interconnection must always be borne in mind by those engaged in forecasting.

Passing Through Bifurcation Points in Noisy Systems. Let r_b denote the governing parameter of a system where bifurcation occurs. Suppose that at $r < r_b$ the system has only only state of equilibrium, $\bar{y}(r)$, while after the bifurcation it acquires two stable states, $\bar{y}_1(r)$ and $\bar{y}_2(r)$. With r changing very slowly, the probability that the system will reach either of these states is $\frac{1}{2}$, since neither of the states is noisier than the other. Therefore, when a bifurcation point is passed through slowly, a new state sets at random. To this it should be added that in approaching a bifurcation point one observes a significant build-up of fluctuations. This phenomenon is of the same origin as the pre-oscillatory noise which strongly amplifies in oscillation sources, specifically, lasers before they start to operate, or noise amplifying at moments when phase transitions take place (a phenomenon resembling critical opalescence). One may think that in traversing a bifurcation point a new state is set under the dominant influence of noise. In actuality, this is not always the case: when r changes rapidly, it is dynamical processes, not noise, that are in control of state selection [7.3]. This is where one of the paradoxes of predictability resides: non-stationary, rapidly varying conditions some-times afford better forecasts than slowly varying, almost stationary ones.

Assume that a bifurcation point is passed through at instant t_0. The governing parameter in the neighborhood of this point can be taken as varying linearly: $r(t) \cong r_b + \dot{r}_b(t - t_b)$, where $\dot{r}_b = dr(t_b)/dt$ is the variation rate at the moment of bifurcation. As things turn out, with $\dot{r}_b$ sufficiently high, $\dot{r}_b \gg G\sigma_f$, the effect of noise can be neglected, and the system has about a 100% chance of reaching a certain steady state, namely, one dictated by equations of motion with a varying parameter $r(t)$. Conversely, at $\dot{r}_b \ll G\sigma_f$, the stationary state is selected exceptionally under the influence of noise, as noted at the beginning of this section.

A critical velocity value $\dot{r}_b \cong G\sigma_f$ proportional to the noise level in system σ_f has been derived both analytically and by numerical modeling. In particular, Butkovsky numerically modeled a logistic mapping that referred to the instant of the first period doubling (the results are to be reported at greater length).

Prediction Costs. However trivial it might sound, prediction costs money. In the area of weather forecasting, the most critical problems are developing networks of observing stations, improvement of data communication, the use of more powerful computers, and so on.

Whenever we embark upon forecasting, constraints imposed by costs eventually have an effect on any factor involved: it is cost that determines the strength of instrument noise (because the instruments employed may be more advanced or less advanced), the amount of initial information underlying the prediction model, or the capabilities of a computer in use.

Prediction cost can therefore be classed as one of the fundamental limits on predictability, and indeed as a new form in which to express the complexity of phenomena under investigation. At any rate, the issue of insufficient investment is mentioned more regularly than those factors which are believed to be 'more fundamental'.

7.6 Dynamical Analogs to Social and Economic Phenomena

The scheme of a forecast based on dynamical equations seems to be promising not only for physical but also for many other phenomena, including those related to economic and social life. Let us consider some aspects of this statement.

First of all, it must be emphasized that far from all social and economic events are of a dynamical nature, that is, capable of being described by dynamical differential equations. Those most closely approaching dynamical analogs are perhaps material production and commodity circulation macrosystems involving large masses of commodities and quantities averaged over both time and territory. The smaller the scale of the process concerned, the greater is the divergence from purely dynamical processes, and hence the smaller the chance to prepare a dynamical forecast. This chance is still smaller for social systems, in which average behavior is often dominated by that of individuals, such as social leaders chosen by other members of the community.

Besides, even in those cases where we can write equations for the social and economic phenomena, it should be borne in mind that those equations have certain distinctive properties, with no analogs to be found in natural science. The fact is that the individual equations and some of their coefficients may vary depending on a host of social and political factors. For example, model equations must, to some extent or other, reflect public psychology (including such extreme phenomena as fanaticism or mass aberration), the extent of trust placed in leaders, dependence on the past and on traditions, and so on. Moreover, coefficients of even the 'most generalized' equations, say, those giving the dynamics of worldwide fuel production, may turn upon decisions made by individual states, or sometimes individual dictators.

Much depends on the laws in force in a given country, on its resources, on the flexibility and sagacity of the leadership, and on people's adherence to dogmas. These factors undermine the predictive power of relevant dynamical models. This certainly does not mean that we should give them up for good, for they can apparently be useful in providing particular scenarios of social and economic development when a serious decision is to be taken.

Though model equations are of limited utility, the dynamical approach to social and economic processes in itself gives rise to truly good analogs with phenomena studied by natural science. For instance, it can be noticed that the transition from one economic and political system to another (now underway in the republics of the former Soviet Union) has much in common with phase transitions. The analogy is to be found in both the structural area (dismantling of some structures and emergence of others) and in some particularities, for example, fluctuation build-up in the reformation period, which may be seen exactly as a phase transition.

In certain circumstances, the dynamical structure of an economy features local instability. This is not a catastrophe for a free market economy; indeed, business adequately responds to perturbations by abandoning unprofitable transactions. On a small scale, instability may cause individual enterprises to disappear, whereas overall stability remains.

Conversely, in centrally planned economies (where low production efficiency is often aggravated by poor management permanently behindhand in adequate decision making) the probability of lapsing into deep chaos grows dramatically. The question of dynamical chaos in economics is becoming an area of intense interest in the literature, and there are even publications addressing the bordering subject of chaos and predictability in national security [7.9].

Generally speaking, the impact of fluctuations in economics seems to be considerable. The so-called major cycles in economic history were presumably initiated by minor impulses (the invention of the steam engine and the power loom, the electric motor, and the computer) that drastically changed methods of production and gave rise to novel technological civilizations in historically short time periods. There are reasons to believe that a healthy economy enjoys the property of local instability with regard to valuable inventions, whereas an unhealthy one just repels them.

Analogs to the above processes that drastically change the 'technological outlook' of economies in relatively short terms (of the order of 50–80 years) can, in all probability, be found in other spheres of life, including geopolitics, demography, ethnogenesis, and so on, although they may involve other and different temporal and spatial scales. Such processes must be more or less predictable during stages of steady growth and saturation rather than at the stage of origination. This bears some resemblance to big air bubbles in turbulently boiling water: as soon as a bubble has become big enough, it continues to grow steadily and predictably on its way upward, but the origin and early growth of a bubble are hard to predict.

As to the 'trajectories' followed by particular enterprises, machines, or men, they can be compared with individual molecules in a boiling liquid: sooner or later, through the action of thermal or hydrodynamic forces, they become involved in the motion of a bubble-to-be, though the latter cannot be specified in advance.

It is all the more impossible to presage the fates of individuals endowed with will and liable to passions, plunged into a social environment and devoted to certain principles. I can only second Andrei Sakharov's opinion on this point, who said: "Destiny is exactly what I don't believe in. I believe that the future is unpredictable and vague. It is framed by all of us, bit by bit, in our infinitely complex interaction."

7.7 Conclusion

Let me summarize the results of the above discussion in a 'tutorial' form so as to emphasize their paucity against the background of truly complicated problems.

1. Any process has a limited predictability range stemming from inaccuracy of the model equation, strength of measurement noise, and fluctuations in the system concerned. The greatest value of the range, achievable in the case where only unavoidable fluctuations within the system itself are to be found, constitutes the predictability horizon, beyond which prediction is impossible.

2. Any means will do in the search for a prediction model, but the greatest success is achieved through the combined use of a good model and a powerful computer.

3. Among many factors limiting predictability there are fundamental ones, namely, local instability of systems under study, the perturbing action of instrumentation, evolutionary processes in predicted systems, bifurcation in the course of evolution, and constraints due to costs.

4. Many notions and models elaborated in natural sciences are likely to be helpful for the prediction of social and economic phenomena, though analogs revealed so far are somewhat superficial and call for closer study.

I take this opportunity to express my sincere gratitude to M. A. Chernavskaya and L. P. Pinskaya for their hospitality and patience that helped me to complete this paper.

References

7.1 Yu.A. Kravtsov: Randomness, determinateness and predictability. Sov. Phys.-Uspekhi 32(5), 434–449 (1989)
7.2 L.Sh. Ilkova, Yu.A. Kravtsov, O.S. Mergelyan, V.S. Etkin: Degree of partial determinateness for dynamic chaos. Izv. VUZ-Radiofizika, 28(7), 929–932 (1985) (in Russian) English translation: Radiophys. Quantum Electron. 28(7) (1985)
7.3 Yu.A. Kravtsov: Randomness and predictability in dynamic chaos. In: A.V. Gaponov-Grekhov, M.I. Rabinovich, Yu. Engelbrecht (eds.): Nonlinear waves, vol. 2: dynamics and evolution. Berlin: Springer-Verlag 1989
7.4 Yu.A. Kravtsov, V.S. Etkin: On the role of fluctuational forces in dynamics of autostochastic systems. Izv. VUZ-Radiofizika 24(8), 992–999 (1981) (in Russian) English translation: Radiophys. Quantum Electron. 24(8) (1981); Degree of dynamic correlation and the problem of revealing of dynamic nature of random processes. Radiotekhtika Elektronika 29(12), 2358–2364 (1984) (in Russian) English translation: Radio Engin. Electron. Phys 29(12) (1984)
7.5 J. Lighthill: The recently recognized failure of predictability in Newtonian dynamics. Proceedings of the Royal Society A407 (1832), 35–50 (1986)
7.6 J.D. Farmer, J. Sidorovich: Predicting chaotic time series. Phys. Rev. Lett. 59(8), 845–848 (1987)
7.7 J.P. Crutchfield, B.S. McNamara: Equation of motion from a data series. Complex systems 1(3), 417–425 (1987)
7.8 L.Sh. Ilkova, Yu.A. Kravtsov, A.S. Pikovskii: Identifying the degree of determinateness of experimentally observed chaotic modes. Sov. Phys.-Lebedev Inst. Reports 11, 7–11 (1986)
7.9 A.M. Saperstein, G. Mayer-Kress: Chaos versus predictability in formulating national strategic security policy. Amer. J. Phys. 57(3), 217–223 (1989)

8. The Future is Foreseeable but not Predictable: The 'Oedipus Effect' in Social Forecasting

I. V. Bestuzhev-Lada

8.1 Historical Background

Forecasting as a scientific technology has a background of many thousands of years. This includes religious eschatology – Hinduism, Buddhism, and Jainism on the one hand and Judaism, Christianity, and Islam on the other; social utopias of all kinds from time immemorial to the present day; the philosophy of history, which incorporates the concepts of cyclical development and of humanity's progress and regression and the clash of Marxism with various non-Marxist attitudes (in the first place positivism) to the question of whether the future is cognizable or not. All basic ideological trends in that prehistory maintained that the future is predictable in principle. That certainty, ingrained in the public mind and still giving quite a few fortune-tellers a chance to earn a living, is shared by virtually all decision-makers, from heads of families to heads of state, in the entire world.

In 1927 the Soviet government asked a group of economists under Bazarov (Rudnev) to predict the USSR's performance in the first five-year development period, 1928–32. Its findings could well have become one of the most significant scientific discoveries of the twentieth century had they been recognized in time. The researchers formulated their conclusion in the following way:

Any long-term economic plan, be it a general plan, a five-year plan, or intermediate (annual) targets, must be a synthesis of directives and prognosis, i.e., it must have a certain target and feasibility studies showing that the goal is achievable. The disputes, so frequent among planners, as to which of the two elements should be given preference, in our opinion make no sense and are due to the fact that the problem is set in the wrong way. Teleology and genetics [goal and trend approaches – I. B.-L.] are not antagonists, but parts of one whole linked in a dialectical way [8.1].

In other words, the authors of the report suggested giving up the futile attempts to make exact forecasts in favor of purely conditional prognoses. Some would project the observed tendencies into the future on the assumption that there is no interference by means of management (genetics) while others would consider the possibility of achieving some desirable changes by means of management (teleology). Both operations would enhance the feasibility of plans and other decisions.

Kant's famous saying could be rephrased in the following way: in long-term planning, teleological targets without genetic feasibility studies are fleshless and genetic studies without targets are blind. An economic plan must be both far-sighted and real or feasible at the same time [8.1].

That article, published in the magazine *Planovoye Khozyaistvo* [Planned economy], was locked away back in the early days of Stalinist rule and remained unknown to readers and of no practical importance until the 1980s.

Thirty years later, in 1957, American experts were asked to investigate the possibility of landing men on the Moon and found themselves in a similar situation. Having no idea of the findings of their Russian colleagues, the authors of the Apollo program arrived at the same conclusions. First, there appeared the same dilemma: either a forecast or a plan. What followed was the idea of technological forecasting, which constitutes the essence of modern scientific forecasting. This idea provides for replacing futile attempts to make unconditional predictions with purely conditional exploratory forecasting based on the analysis of trends (like the genetic prognosis made by the Russians) in order to detect or specify problems to be resolved by means of management, on the one hand, and with conditional normative forecasting (which the Russians called 'teleological'), based on the analysis of counter-trends, extending from the tree of goals according to preset criteria and in fact representing the optimization of the trends observed, on the other. The aim of research is the same: to make a preliminary inquiry into the expected and desirable effects of plans, programs, projects, and day-to-day decisions.

This time the discovery was not locked away for political reasons, but was used in practice and had a colossal economic result (up to 800% profit resulted from the optimization of decisions following the comparison of exploratory and normative forecasts), which brought about a 'prognosis boom' in the 1960s. Both in the West and, in the second half of the decade, in the USSR and other East European countries, there appeared hundreds of research centers of varying calibre invloved in developing 'technological forecasting'.

However, it is quite significant that psychologically the popular predilection for prognoses that look rather like predictions (expressed in the public mind as a very strong desire to know one's future, although this is illogical and unnatural, because it makes the life of any living being, including humans, meaningless – we discuss this in greater detail later) has proved so strong that technological forecasting proper has made itself felt primarily at the level of applied science and engineering, where inquiries into the propects for the development of 'man–technology' systems and their subsystems were required.

Meanwhile, in social forecasting – from the future of humanity in general to special social or political forecasts – either the 'teleological' approach or the 'genetic' one continued to dominate. The effects of that were disastrous. In the West it led to the emergence of purely utopian ideas of a 'post-industrial society' emerging from the current 'industrialized society' without political struggle (whose results can never be foretold with 100% certainty) in the 1960s, to be followed by extremely extrapolated concepts of an impending 'global disaster' independent of the unpredictable results of political struggle in the 1970s. As a result, when the ideology of neo-conservatism emerged the winner in the 1980s, Western futurology was plunged into a deep crisis that has persisted ever since.

In the Soviet Union 'technological forecasting' developed rather quickly from 1967 to 1970, in vigorous confrontation with dogmatism in the wake of the failure of Krushchev's plan to build communism by 1980, only to suffer a crushing defeat in 1971 when Kosygin's reforms of 1966–1969 were curtailed. It was replaced with the 'Comprehensive program for scientific–technological progress' extending until 1990 (it was prolonged every five years), which was neither a prognosis nor a program or plan, and consisted of time-serving directives expected to 'support' planning. The planners themselves ignored it. That mockery of planning (the plans had nothing to do with reality) had behind it a peculiar cross-breed of quasi-prognoses and pseudo-programs having purely ideological functions, but kept secret by virtue of their utter inconsistency. Today all this has been abolished as having nothing to do with science or politics, but not yet replaced with anything constructive, because ways of studying the future (an understanding of this has just begun to establish itself in official terminology) remain extremely vital to Russia, as well as to all the other ex-socialist countries.

The problem of the limits and parameters of predictability in social sciences stands tall. Back in the 1960s the problem was often confined to what was called the 'Oedipus effect' – the name was borrowed from the legendary Greek king who tried, in vain, to change his future as revealed to him.

8.2 The 'Oedipus Effect' in Social Forecasting

The term 'social forecasting' has several meanings. In the most narrow sense it denotes the prognosis of social relations that go beyond the bounds of economic, political, and other relations. In the broad sense it refers to prognoses in social sciences, in contrast to prognoses in natural sciences. There are some intermediate meanings, too. Here this term is used in the broad sense, to include economics, politics, psychology, pedagogy, law, and other sciences.

Everything that is subject to forecasting can be positioned on a certain scale showing the interaction of a prediction with the decisions and actions that can change the prediction. At one end of the scale can be found objects with a very small degree of such interaction (close to 0), while the objects at the other end are extremely vulnerable to change and the degree of interaction is rather significant (close to 1). In the latter case the decision can interfere, compensate for, or even refute the prediction. At one end we will find subjects treated in natural sciences, and at the other end those treated in social sciences.

Theoretically, the degree of interaction cannot be equal to zero in either case. For instance, theoretically, one can imagine interference even in the case of a prediction of the eclipse of the Sun, generally regarded as a paradigm of precision. For that it is enough to make a big enough nuclear explosion on the Moon to deviate it from its path by several nanometers (this is a purely speculative example, of course). In application to natural sciences, including technology, we can do very little to change predictions. Weather forecasting is an example. True, we have learned how to change weather in a limited region, but only to a very small degree. Besides, this is extremely costly and very ineffective. For that reason, the sole option left in practice is to try to make predictions and then adjust onself to them. This situation is very frequent in the natural sciences and engineering.

In social sciences, decisions and predictions interact in a totally different way. Individual behavior or political struggle are examples. Whatever I tell you will happen tomorrow or next year on the basis of a thorough study of your character, mode of life, and external factors, as soon as you hear my prediction you will be able to ensure that it never comes true. Or, on the other hand, it may have never occurred to you that anything like that could happen, but once you know the prediction you can translate it into reality. Likewise, in politics it is quite possible to predict a government coup by an army general that is sure to take place the day after tomorrow. But it will then be very easy to dismiss the general and nip the coup in the bud on the basis of the same prediction. Alternatively, it is possible to fan government coup hysteria to a point where the dullest general will have the idea of translating the prediction into reality.

In modern forecasting this is called the 'Oedipus effect'. Detailed descriptions of this phenomenon may be found in the specialist literature of the 1960s. In Russian, a fundamental description of the effect is contained in a monograph by Gendin [8.3]. Nevertheless, clients have as a rule continued to require that a prediction should meet their expectations, and researchers in turn have pretended that there is no Oedipus effect.

To get over the psychological barrier that blocks the way from fortune-telling to prognoses evaluating the effects of likely decisions, one has to address onself to this question over and over again. In this chapter, to build up from the basic methodological ideas of modern scientific prognostication we suggest:

Postulate 8.1. *In principle it is impossible to predict with 100% certainty the future states of objects that can be changed by actions taken on the basis of decisions stemming from the prediction.*

This postulate is in turn derived from the postulate that in principle it is impossible to predict the future state of any organism capable of responding to such a prediction by changing its behavior, because the knowledge of the future in this situation makes the life of the organism meaningless and is therefore physically impossible.

Indeed, imagine that some fish knows for sure that it will open its mouth in the morning to swallow a worm under any circumstance and that it will open its mouth in the evening of the same day only to be hooked, also under any circumstance. It is clear that the fish will sooner or later be replaced by some cybernetic organism having nothing in common with the Earth's animal or plant life and still less with the laws of existence in our domain of the universe.

Similarly, it is hard to think of a human being who knows with certainty that he or she will have a merry wedding tomorrow only to die in an accident the day after tomorrow. Clearly, such people find themselves in a situation – described so many times in fiction, from Aleksander Pushkin's *Egyptian Nights* to Mikhail Bulgakov's *Master and Margaret* – where they either humbly wait for the prediction to materialize or try to avoid it, or turn into supernatural beings, alien to anything human. In this connection it is worth recalling one example, from *Master and Margaret,* to illustrate the thesis: from the human point of view the mere existence of Voland and his associates, who knew in advance what would happen to them in the book and beyond its bounds, was utterly meaningless!

From that point of view, the foreseeing of the future (that the future is foreseeable is a hard fact supported by quite a few historical developments) can in practice take the form either of adjustment to a 100% certain prediction (which cannot be changed physically) or the analysis of trends mentally extended into the future. The laws by which those trends developed in the past and develop in the present are well known, including the detection of long-term problems to be resolved with management means in case the unfavorable trends persist, and the optimization of trends to highlight likely solutions to problems detected as a result of prognostic exploration.

If Postulate 8.1 (will it ever be of any use?) is employed to systematically retarget prognostic inquiries into the behavior of objects that are manageable in principle, the effectiveness of such prognoses can grow tangibly.

To simplify the explanation we deliberately ignored two important ideas. Firstly, apart from the extreme readings (0 and 1) on the scale mentioned earlier, there is an infinite number of intermediate readings true of the cases where the interaction of the prediction and the decision is partial and takes place only under certain circumstances. Secondly, the interaction of the prediction and the decision that is possible *in principle* does not imply that this possibility is bound to materialize. This holds not only in those cases where we are 'out of reach' but even when very little is required to change what has been predicted. Just consider how many things that can be controlled in the most rational way we – humanity – let develop as they do, in the same way as spontaneous processes in nature. This applies not only to natural phenomena or technology but also to any development in the social sphere – from the economy, culture, and politics to intimate family affairs.

Have we learned to control demographic processes? Far from it. They are still in many respects similar to spontaneous climatic and weather changes. This is true, to a certain degree, of any process or phenomenon being studied by social scientists. For good reason they could be called spontaneous, or natural. However, since they are controllable in principle, the established term for them in modern forecasting is quasi-natural, or quasi-spontaneous. This quasi-spontaneity does no good to people themselves. Researchers seem to be the sole exception, because it makes it possible for them to minimize the Oedipus effect, to ignore it and thereby make forecasting far simpler.

This allows us to formulate:

Postulate 8.2. *The quasi-spontaneity of most social processes or phenomena, i.e., the uncontrolled occurrence of those that are controllable in principle, is reason enough to regard them in a sense as natural phenomena and to ignore, to a certain degree, the fundamental possibility of controlling them. This considerably facilitates and simplifies the analysis of trends, as well as their optimization, and taps the tremendous potential of scientific forecasting, whose aim is not to make indisputable predictions but to raise the effectiveness of decisions by making purely conditional predictions that preview the effects of likely decisions.*

In practical terms, the point at issue is the fundamental possibility of dividing the process of future research into two stages: abstraction from likely decisions capable of altering the tendencies observed (exploratory forecasting) and abstraction from likely decisions capable of interfering with the optimization of trends (normative forecasting). The mandatory distinction between two stages of forecasting and the equally mandatory comparison of results makes it possible to avoid any bias, either utopian (apologetic, justifying current policies) or anti-utopian (alarmist, critical of current policies). In other words, this makes it possible to achieve a far greater impartiality and effectiveness of forecasts (in terms of feasibility) and to resist the temptation to follow in the footsteps of the notorious authors of bankrupt predictions.

Let us try to illustrate this with two examples. One concerns globalistics – a set of contemporary global problems – and the other concerns the outlook for *perestroika* in the Soviet Union (ignoring specific regional issues that cannot be taken into account here).

8.3 The Problem of Foresight and Prediction in Globalistics

Gone are the days when the West described humanity's foreseeable future as the transition of industrialized countries to the idyll of a 'post-industrial society' that was expected to do away automatically with such global problems as the widening gap between countries at fundamentally different levels of development; the aggravating imbalances in the supply of energy, raw materials, and foodstuffs, in patterns of trade, transport, communications, ecology and demography, in the military sphere (the arms race), and others; the crises of world culture, education systems and health service; the nightmares of hyper-urbanization on the one hand and the degradation of villages on the other; the tides of antisocial phenomena such as crime, the 'shadow economy', drugs, and alcoholism; and the inability of international organizations to do anything about them all. Likewise, there can be no return to the alarmist pictures of an impending global disaster that may occur if only the above problems acquire critical dimensions: people's minds get used to the attempts to fan such fears, then become tired and ultimately want to hear something new.

It became quite clear more than a decade ago that the current world civilization is unlikely to survive the 21st century. To be able to live on in the current conditions a civilization will have to be dramatically new, consume less energy, and be more stable in ecological, demographic and cultural terms. Ways of achieving that are more or less clear: an end to industrialization, urbanization, and mass cargo and passenger traffic using any motor transport; measures to increase in every possible way the share of ecologically 'clean' sources of power (solar energy, water, wind, the heat of the Earth's interior, and the difference of temperature between upper and lower layers of ocean and sea water, etc.) in the world energy balance; recycling of wastes and reusable materials; the achievement of a demographic optimum by all countries of the world – three children on average per woman of reproductive age; the optimization of education so as to eliminate the gap between generations; the stabilization of culture by eradicating the destructive addiction to modern *ersatz* culture; the minimization of phenomena detrimental to society, such as crime and drug abuse; and the priority of such basic values as the family, science, the arts, and meaningful pastimes; and – in first place – the comprehensive development of the individual. However, the psychological hurdles blocking the way to an alternative civilization are rather high. Besides, there are also social, economic, and political barriers (economic backwardness, traditions, political struggle, etc.). The past twenty years of attempts to do away with them have ended in failure. Apparently, efforts along these lines should be increased tenfold, including those in the field of forecasting.

Just recently, in the 1960s and 1970s, when the contemporary theory of forecasting was just finding its feet, a rational stratification of conditional technological prognoses into the following five levels emerged: current (minor quantitative changes), short term (significant quantitative changes), medium term (quantitative–qualitative changes), long term (qualitative–quantitative changes), and extra long term (general qualitative changes). The length of the period previewed was set in different ways, depending on the subject of the prognosis. In the Soviet Union, the current period for social and economic prognoses was confined to one year or the forthcoming season, the short term to the forthcoming five-year plan, the medium term to the subsequent five-year plan, the long term to one or two subsequent five-year periods (but no more than 20 years), and the extra long term extended beyond that limit.

Over the past ten years, socio-economic and socio-political changes in the world have quickened their pace still further. Long-term prognoses for 20 years in advance are no longer relevant. Long-term prognoses these days have been reduced by no less than 30% and fit perfectly into the last five years of the 20th century. Accordingly, the range of the medium term has shortened to 3–5 years and that of the short term to one year, and current prognoses have increasingly often been made not so much for months as for weeks and even days ahead. The next century lies within the scope of extra-long-term forecasts, which now ever more insistently require stratification. As a matter of fact, this work lies far ahead.

Additionally, there has been a growing fundamental difference between long-term and extra-long-term prognoses: the former preview qualitative–quantitative parameters of the processes and phenomena in question whereas the latter are confined primarily to general qualitative evaluations and tentative quantitative illustrations. Everything beyond the next century looks as vague and uncertain in terms of forecasting as the 21st century did in the early days of modern forecasting theory – a mere 25 years ago.

All qualitative prognoses for the 21st century, without exception, are very conditional and uncertain and have nothing to do with predictions proper. Even highly inertial social processes, such as the growth of world population, are impossible to foresee with a high degree of accuracy. True, it has been computed that in the first quarter of the next century the world's population will be up from today's five billion or more, and will almost certainly rise from more than six billion in 2000 to eight billion or more, then in the next 50–100 years will stabilize at a level of ten billion (provided the fall of birth rates continues). The error range constitutes two or three billion. But is this what one can call a prognosis? Moreover, one must bear in mind that we have ignored the high probability of disasters, such as devastating wars and AIDS on the one hand, and the seizure of power in a big country (or countries) in the Third World by clerical–fascist forces, capable of slowing down the falling birth rates and doubling the population over a mere 50 years, on the other. In either case the error range may be as wide as plus or minus five billion, which is as far from an accurate prognosis as the weather-man's forecast of 'either fair or cloudy'.

Processes featuring a high degree of discreteness (and they are in a majority) are still less predictable. What can one say about the future of the world energy balance except to assign a certain probability to the achievement of controlled thermonuclear fusion, which is capable of changing this balance qualitatively? Only one thing: in the 21st century it will no longer be possible to double the production and consumption of energy, something humanity has done in the 20th century, the more so since in the next century the Third World countries will contribute to the energy balance on a far greater scale than now – and they will account for nine-tenths of the world population in 2100. Indeed, this prognosis has tremendous importance for global, regional, and even local political decisions.

What can one say about the outlook for the information revolution, which is paving the way for information exchanges in defiance of time or distance? Here one must also remember the possibility of sublimating society's transport requirements and replacing them with those for electronic communication. At least it is time to start seriously analyzing possibilities of this kind, which are about to change fundamentally not only the lifestyles but also the mentality – or in any case the social psychology – of *Homo sapiens,* even if it is not possible to make definite predictions.

All of the basic parameters of the possible and necessary alternative civilization of the 21st century are unpredictable in principle and purely normative: to survive the 21st century the human race *must* achieve a demographic optimum of three children per woman, it *must* minimize the use of motor transport and maximize waste-free manufacturing, it *must* minimize the catastrophic generation gap and the disastrous effects of *ersatz* culture on the individual and on society, and so on. This takes us back to the historic significance of the discovery of the possibility in principle of replacing futile attempts to predict controllable processes with technological prognoses, in the first place aimed at making control more effective.

The above qualification is applicable, to a lesser but still significant degree, to the 1990s. True, inertial processes do expand the horizons of prognoses and predictions. For instance, no conceivable force can alter the current trend that will increase the world's population from five billion to six billion over the next ten years, except for nuclear Armageddon (which can result from an improbable provocation or failure in technical systems) or other global disasters of a kind similar to natural disasters. In turn, the range of such processes – contrary to that of discrete ones – is rather limited, and beyond it there lies the domain of unpredictability.

What can one say about world oil prices – a very typical discrete phenomenon? Any computations can be shattered by a conflict, capable of pushing oil prices up or down 50% and more, a conflict that is largely a result of the personal character of this or that statesman. What kind of relations will Iran, Iraq, Kuwait, the United Arab Emirates, Saudi Arabia, and other Arab countries have in the 1990s? [This chapter was written before the 1991 Gulf War – I. B.-L.] No predictions can be made. One can either guess, with little chance of hitting the nail on the head, or make exploratory or normative forecasts to optimize the foreign policy decisions of this or that government. This is so in the overwhelming majority of cases, in all fields of social (socio-economic, socio-political, etc.) forecasting.

8.4 The Problem of Foreseeing and Predicting the Development of the Former Soviet Society

The aforesaid has a direct bearing on prognoses made on a regional scale (in contrast to the global scale), including forecasting the development of the former USSR. The smaller and less complicated the subject of a forecast, and the fewer the trends counterbalancing each other, the narrower is the range of predictability, although future reseach may seem rather simple in such a case.

Was it possible in the first half of the 1980s to predict the *perestroika* that would occur in the second half of the decade? True, reform was becoming a growing necessity, and one could predict with a rather high degree of certainty that it would take place within one or two decades, three at most, the more so since it would be the sixth attempt at reform. Lenin's new economic policy of 1921–29 (in practice it was over in 1927) was the first, Krushchev's reforms of 1956–64 (in practical terms they ended in 1962) the second, Kosygin's reforms of 1966–71 (practically of 1966–68) the third, the draft reforms of 1979 that never materialized the fourth, and Andropov's 1983 attempt to put an end to stagnation the fifth. All were attempts to breathe new life into the utopia of barracks socialism (the system of management by decree, burocratic centralism, totalitarian rule, etc.) and all failed, because an utterly inefficient system proved amazingly viable. Cemented by the world's most powerful burocracy relying on the 'shadow economy', the system suppressed the people by wholesale terror and brainwashing. It distorted the morality and mentality of people at the individual level. A simple historical analogy was enough to predict another inevitable attempt to overhaul the system in the near future and the high probability of another failure. However, in practical terms a prediction like that was very difficult to make.

Let us bear in mind that the upper echelons of the system, scared by Krushchev's arbitrary policies, had taken every conceivable precaution to avoid a repetition of anything like the Krushchev thaw. Even Andropov's attempt to revitalize Soviet life was given the cold shoulder. The members of the country's real government, the Politburo of the CPSU Central Committee, were chosen on the basis of their personal credibility (though formally they were elected by the CPSU Central Committee at its full-scale meetings), and any deviations in character from the norm were counterbalanced by the close connection of the ruling clans, who were supported by a solid basis of conservative first secretaries of regional party committees and heads of crucial ministries and agencies – those who were in the commanding position locally. In a situation like this, Mikhail Gorbachev, a protegé of Andrei Gromyko and Yuri Andropov, would have no chance of becoming a candidate member of the Politburo, let alone General Secretary, had he just once taken the liberty of saying something, however casual and informal, that would give reason to suspect him of reformist intentions. Gorbachev was elected General Secretary for the sole reason that he was one of the youngest, healthiest, and most active members of the Politburo amid the confusion following the deaths of three General Secretaries over a period of two years. That factor made the well-oiled machinery of 'succession to the throne' on account of the candidate's proximity to the previous General Secretary laughable in the eyes of the whole world and blocked most of Gorbachev's contenders from taking the highest post. Yet, even in that situation the new General Secretary was elected by a very narrow margin. Intrigues in the party apparatus could have had a totally different net effect.

Aware of these circumstances, one could expect with great certainty another attempt at reforms in the 1990s, with an error range of five years, probably similar to the 1979 plan for face-lifting reforms, or a milder version of Andropov's efforts at best. Neither option would infringe on the system of one-party rule or the system as such. Gorbachev looked like a 100% party burocrat, with the background of a career regional Communist Party official. However, he has proved not very typical, or rather, by virtue of events, very untypical. Yet whatever he is, his practical potential as a reformer is rather limited not only by his character (which has displayed an amazing ability to develop – in contrast to hardline Communists) but also, in the first place, by his political environment. His original environment was conservative. Nonetheless, the team he later chose himself proved still worse. Except for two or three figures all were militant conservatives. This explains why the policy of *perestroika* stumbled now and then throughout the period 1985–90. The slow-going steam engine of Gorbachev's reforms took the daily risk of becoming detached from the outdated and rattling wagons of party burocracy. If that happened, dire consequences were bound to ensue. Considering the 'ifs' of the past is beyond the bounds of modern science. True, this does not rule out exploratory or normative predictions – but who would dare make any just a year or two ago, let alone in earlier days? The existing social psychology simply left no room for them.

Can one assess the probability of any of the outcomes of *perestroika* that have been openly discussed in the West for the past 3–5 years? Let us imagine that the situation of the second half of the 1980s has continued to this day and will remain unchanged until the end of the 1990s and even into the next decade – the global situation will simply not allow us to look any further. Let us imagine that a balance between supporters and opponents of *perestroika,* preventing it from gaining strength but giving no chance of curtailing it either, together with the slide in the economy and politics from crisis to disaster, the weakening of Communist Party rule, and the slow emergence of democracy continue throughout that period. In that case, the USSR will find itself in a position similar to that of India, Brazil, or Mexico. Its future will be decided by external or centrifugal forces beyond the confines of the Kremlin. Or let us take a 'Marshall Plan' for the Soviet Union, which would put the country on the track of a market economy and help it follow in the footsteps of Germany or Japan to join the club of the world's industrialized nations (such precedents are rather frequent these days). Neither should one rule out the possibility of a military coup restoring a neo-Stalinist regime – bearing in mind those who are at the head of the army and the KGB security police it is hard to think of any other option. Or the Soviet Union may decay into totally autonomous regions – the process is going on as fast as the deployment of armed forces will permit it. Or the crisis may be ended by a series of sudden and highly effective socio-political and socio-economic decisions by the government – although the chances for this look bleak, primarily because of the rivalry mentioned above between reformist and conservative forces. Any other conceivable scenario can prolong this list.

We believe that the methodological idea presented earlier that controllable processes and phenomena are unpredictable in principle is fully applicable to the whole range of these possibilities. True, given the appropriate information, we could predict the date and form of a Soviet analogue of the 'Marshall Plan', should such a document appear in sight; it would be possible to foretell the date and nature of a government coup, or the separation of certain regions from the country, or an effective government solution of the whole range of reform-related problems (the author knows nothing of any such solution). But that would mean an upsurge in the struggle of supporters and opponents of *perestroika* over any 'Marshall Plan'-type option. The outcome of this struggle would either accelerate or delay such a dramatic step, and besides, could seriously change the form and content of that measure in either direction. The same is true of the prediction of a military coup, the decay of the empire, or an effective solution of the problem. In other words, instead of scientific prognostication we will find ourselves with sheer propaganda. The consequences will be quick to ensue.

This has a direct bearing on all the other problems society is likely to encounter, be it the restructuring of the economy, the fate of party burocrats, the democratization of society, the return to stability of two conflicting demographic trends (the demographic explosion in Central Asia that is likely to double the population there over a period of 20–30 years versus the depopulation of all other regions of the country), the modernization of inadequate education and health service systems, resistance to the disastrous decay of culture, hyper-urbanization and the degradation of rural areas, the destruction of the environment, the onslaught of crime (including latent crime in the 'shadow economy'), the fight against heavy drinking coupled with effective measures to minimize the effects of a massive drug invasion, and so on.

However, technological social forecasting will acquire ever greater importance, for it allows the evaluation (through the comparison of exploratory and normative conditional prognoses) of the likely consequences of decisions that are already made or still being contemplated concerning each of the mentioned problems, and thereby makes these decisions more precise and effective.

The specific traits and nature of conditionality in both exploratory and normative prognoses constitute an independent scientific problem, which requires a special discussion.

References

8.1   Planovoye Khozyaistvo [Planned economy] 2, 38–39 (1928) (in Russian)
8.2   Kakim Byt Planu: Diskusii 20-kh Godov [What should the plan be like: discussion of the 1920s], p. 166. Leningrad: Lenizdat Publishers 1989 (in Russian)
8.3   A.M. Gendin: Predvideniye i Tsel Razvitiya Obshchestva [Foresight and aim in society's development], pp. 236–290. Krasnoyarsk 1970 (in Russian)

Appendix A: Looking Back on the August 1991 Coup

Who would have dared predict in detail the events that took place in Kuwait or in Moscow in 1990/91? Nevertheless, the forecast of the march of events in the USSR made by several research groups independently of each other at the end of 1987 and the beginning of 1988 has come true in almost every detail. Besides, what happened in this country and elsewhere in the summer and autumn of 1991 has convincingly shown the effectiveness of the main postulate of technological forecasting.

Let us recall that the policy of perestroika, launched in 1985, according to those forecasts, could go on as it did for no more than five years. That would inevitably be followed by a qualitatively new scenario.

So it happened. By the autumn of 1990, five years after it had begun, perestroika reached a dead end and the transition to 'Scenario 2' – a military dictatorship – began. Several months were taken by efforts to find an equivalent of the Marshall Plan for the USSR (Scenario 3) as the sole realistic opportunity of ending the crisis.

August 1991 saw the comeback of Scenario 2, followed by another turn to Scenario 3; at that time Scenario 4 providing for the USSR's decay into independent regions was already much in sight. Scenario 5 remained the least probable, just as stated in 1988; it would place power in the hands of a charismatic leader, who would lead all member-republics of the now-dissolved USSR out of the crisis at the expense of great, though temporary, sacrifices, as happened in Poland.

As we have seen, technological forecasting, although it can never be 100% certain, proves very helpful in drawing an impartial picture of likely events and assessing the degree of their probability.

December 1991

Appendix B: Looking Ahead

The future of Russia and the other former Soviet republics looks still less predictable now, another year later, at the end of 1992.

It has proved that translating a utopia into reality is much more difficult than doing away with it. Even the Eastern *Länder* of Germany (formerly the German Democratic Republic) will need as long as a whole generation, according to some experts, to become fully integrated with the Western ones. The same applies to the former Soviet republics to a much greater degree.

The biggest and most unpleasant discovery of 1992 was that once industrial recession and the collapse of the state have begun, it is very difficult to stop them. No less than several years have to go by before the frustrated economic ties of a command economy established from 'above' can be compensated for by market economic ties emerging from 'below'. It will take years to make millions of people, who have got used to simulating work in exchange for simulated pay, start doing their jobs in earnest as is required by a market economy, and first they have to go through mass unemployment. It will take years for the centrifugal trends of 'breaking away from Moscow' to be overpowered by the forces of economic cooperation and political integration, just as it did in Western Europe. Experts claim that no less than 10–15 years will be needed to end the period of recession and stabilize the situation at least at the level of 1985 (in a different economic and political dimension, of course). After that, in the middle of the first decade of the next century, one may expect an upturn.

Since many former 'socialist' countries have already gone far ahead of Russia, one can employ the method of forecasting by analogy to say that by 2000–2005 – on the condition that optimal decisions are made all the way – Russia may achieve Poland's 1992 level (an average monthly wage of US$ 150 instead of Russia's US$ 15 at world prices, but still without full adaptation to them). It will take at least as much time again to achieve Turkey's 1992 level (US$ 300 a month and full adaptation to the world market).

This is how the foreseeable future of Russia and the other former Soviet republics appears. It can be drastically worsened by inadequate decisions to an utter economic collapse and civil war on the Yugoslav model. But it will be impossible to bring about anything similar to the

Korean or Taiwanese miracles – the forces of retardation are too great. Incidentally, Korea, Taiwan, and the other Far Eastern 'tigers' also needed quite a few years to accomplish their miracles.

As far as predictions are concerned, everything looks far more difficult. Anything can happen tomorrow: a military coup, a return to the past, or fierce hostilities, like in Yugoslavia. However, that would only slow down the inevitable movement away from a social pathology (totalitarianism) to the social norm (democracy and a market economy). Of course, that norm should not be idealized. There one finds different types of problem; however, these problems are 'normal' and not 'pathological'.

The sole constructive approach to such a situation is to turn for help to the theory of probability, just as in the case of 1987/88. This approach makes it possible to say that most likely (with 60% probability) the process described above will go on as expected, albeit with various complications, for the next 20–30 years. It is less probable (say 30%) that it will be complicated by a temporary return to totalitarianism and/or civil war. It is still less probable (10%) that an adequate equivalent of the Marshall Plan for the former USSR will be devised and that Western technologies and investments will help the victims of totalitarianism regain their feet in less than 10–15 years, say in 3–5 years. And it is most unlikely (a tiny fraction of 1%) that some new charismatic leader – there have been none in sight so far – will be able to mobilize the nation and make it catch up with Poland in a leap, within a year or two at the latest.

But is this information redundant in making optimal decisions?

December 1992

9. The Self-Organization of American Society in Presidential and Senatorial Elections

V. I. Keilis-Borok and A. J. Lichtman

9.1 Historical Background

The American electorate is commonly regarded as a hierarchic system of electoral groups that differ in their responses to the issues and tactics of electoral campaigns. Taken together, the groups comprise the entire electorate; they are normally represented in social or political organizations, but do not coincide with them. Cultural, social, economic, territorial, ethnic, and other factors divide voters into electoral groups. Each group, in turn, can be subdivided into smaller units, down to the indivisible, though variable element of the individual voter.

American elections are often summarized as follows:

- The candidate's task is to optimize his platform and campaign strategy to win over as many electoral groups as possible, without antagonizing too many other groups.
- At the peak of electoral campaigns, the quest for such optimization goes out of proportion, which makes the campaign shallow and separated from the real interests of society.
- The outcome of the election to a large extent depends on the ability of each candidate to manipulate electoral groups through advertising, rhetoric, and stage-managed events.

This formulation appears plausible, and it has consistently guided coverage of American elections. But it leaves a number of questions unanswered. Though bifurcations do occur in nature, there needs to be specific evidence for the belief that a huge electorate (10^8 for a presidential election and 10^6–10^7 for senatorial ones) easily changes its collective decision under the influence of transient factors that are irrelevant to the governing of the country.

Besides, this account shows little respect for American democracy or the American voter. It portrays voters as excitable simpletons who are easily swayed by television commercials and emotional appeals.

The problem of elections in the U.S. deserves an alternative hypothesis. In our papers [9.3, 9.5, 9.6] we have proposed the hypothesis that the collective decision of the American electorate (on the national scale when the president is elected, or in each state during senatorial elections) is determined by long-term factors that can usually be gauged well in advance of an electoral campaign. This means, first, that debates, advertising, nightly news reports, and other campaign events have little to do with the outcomes of elections and, second, that election results can be predicted reliably many months in advance without recourse to polls.

In [9.5] we have tried to find parameters which describe the political environment prior to an election and predict the outcome of American presidential elections. In [9.6] we examine the same problem for senatorial elections in all 50 American states.

Given a lack of adequate theory, both problems were tackled phenomenologically through the analysis of pre-election situations in the past. The results are relevant for understanding the electoral process, developing theoretical models, and for predicting electoral results.

Our work shows that presidential elections are referenda on the performance of the party in power. Rather than choosing between individual candidates or between Republicans and Democrats, the electorate decides whether to retain or reject the party that controls the White House. If the nation has fared well during the term, the executive party is re-elected; otherwise the presidency changes hands. This pattern has remained unchanged at least since the election of Abraham Lincoln in 1860 despite enormous changes in the electorate and the society during the subsequent 130 years.

Although other investigators have also probed the conditions under which incumbent administrations rise or fall, their work has focused narrowly on the state of the economy. Strictly economic models, however, fail to capture the dynamics of many American elections (e.g., 1860, 1912, 1968, and 1976). Our work in [9.5] gauges not just the economy, but such measures of incumbent performance as policy change, foreign successes and failures, scandal, and social unrest.

We found that the choice of a senator is fundamentally different from the election of a president. Senate elections are not primarily referenda on the performance of the previous term. They are to some extent contests between parties and candidates, although in fact their outcome depends more upon the relative strengths of the competitors going into the campaign than on what they say or do during the campaign itself. That is why we found that senatorial elections, like presidential elections, could also be forecast in advance of campaigns.

The logic and algorithms of the present analysis follow Gelfand's school of pattern recognition analysis. Specifically, we draw upon the experience of earthquake prediction research [9.1]. For both presidential and senatorial elections we sought to predict the winners of elections, but not their percentage of the vote. Likewise, we eliminated certain information through the discretization of parameters to the lowest level of resolution, 'yes' or 'no'. Similar 'robust' methods are widely used in the heuristic analysis of complicated data, especially when dealing with small samples. The apparent loss of information involved has often produced the stable results that elude more 'detailed' analyses that are subject to fluctuations in the values of particular variables.

9.2 The American Presidential Election: Formal Analysis

Initial Data. The historical study of American elections suggested the choice of a set of integral parameters that may affect the outcome of presidential elections. We were guided by the hypothesis that elections turned on the strength and performance of the incumbent party. These parameters were determined at the minimum level of resolution, as answers to certain 'yes' or 'no' questions given in Table 9.1. Each question can be answered prior to an upcoming election, usually by the time that parties have nominated their candidates. The questions are formulated so that an affirmative answer favors victory of the party in power. The unexpected finding is that this set of questions is sufficient to predict the outcome of elections.

The answers to the questions for pre-election situations from 1860–1980 are given in Table 9.2. This material forms our initial data.

The analysis follows a procedure that is common for pattern recognition. Let us divide the pre-election situations into two types: *I* (incumbent will win) and *C* (challenger will win). Victory is defined as a popular vote plurality regardless of the verdict in the electoral college.

The problem is formulated as follows. We are given 'learning material' that consists of examples of elections *I* and *C*, with the answers to the questionnaire (Table 9.1) for each election. From this data

Table 9.1. Questionnaire for presidential elections

A 'yes' answer favors the incumbent party. When five or more answers are 'yes' a victory for the incumbent party is predicted. Otherwise a victory for the challenging party is predicted.

1   **Party mandate.** After the midterm elections, the incumbent party holds more seats in the U.S. House of Representatives than it did after the midterm elections.

2   **Contest.** There is no serious contest for the incumbent party nomination.

3   **Incumbency.** The incumbent party candidate is the sitting President.

4   **Third party.** There is no significant third party or independent campaign.

5   **Short-term economy.** The economy is not in recession during the election campaign.

6   **Long-term economy.** Real per-capita economic growth during the term equals or exceeds mean growth during the previous two terms.

7   **Policy change.** The incumbent administration effects major changes in national policy.

8   **Social unrest.** There is no sustained social unrest during the term.

9   **Scandal.** The incumbent administration is untainted by major scandal.

10   **Foreign military failure.** The incumbent administration suffers no major failure in foreign military affairs.

11   **Foreign military success.** The incumbent administration achieves a major success in foreign or military affairs.

12   **Incumbent charisma.** The incumbent candidate is charismatic or a national hero.

13   **Challenger charisma.** The challenger party candidate is not charismatic or a national hero.

Table 9.2. Presidential elections 1860–1988 classified by questionnaire

Question numbers 1–13 correspond to Table 9.1. Answer 'yes' = 0 and 'no' = 1.

Incumbent victories

Year	1	2	3	4	5	6	7	8	9	10	11	12	13	Total
1864	0	0	0	0	0	1	0	1	0	0	0	1	0	3
1868	0	0	1	0	0	0	0	1	0	0	0	0	0	2
1872	1	0	0	0	0	0	1	1	0	0	0	0	0	3
1880	0	1	1	0	0	0	0	0	0	0	1	1	0	4
1888	1	0	0	0	0	0	1	1	0	0	1	1	0	5*
1900	1	0	0	0	0	0	0	0	0	0	0	1	1	3
1904	0	0	0	0	0	0	0	0	0	0	0	0	0	0
1908	0	0	1	0	0	1	0	0	0	0	0	1	0	3
1916	1	0	0	0	0	1	0	0	0	0	0	1	0	3
1924	1	0	0	1	0	0	0	0	1	0	0	1	0	4
1928	0	0	1	0	0	0	1	0	0	0	0	1	0	3
1936	0	0	0	0	0	0	0	0	0	0	1	0	0	1
1940	1	0	0	0	0	0	0	0	0	0	1	0	0	2
1944	1	0	0	0	0	0	0	0	0	1	0	0	0	2
1948	1	0	0	1	0	1	0	0	0	1	0	1	0	5
1956	0	0	0	0	0	0	1	0	0	0	0	0	0	1
1964	1	0	0	0	0	0	0	0	0	1	0	1	0	3
1972	1	0	0	0	0	1	1	0	0	0	0	1	0	4
1984	0	0	0	0	0	1	0	0	0	0	1	0	0	2
1988	0	0	1	0	0	0	1	0	0	0	0	1	0	3

Challenger victories

Year	1	2	3	4	5	6	7	8	9	10	11	12	13	Total
1860	0	1	1	1	0	0	1	1	0	0	1	1	0	7
1876	1	1	1	0	1	1	1	0	1	0	1	1	0	9*
1884	1	1	1	0	1	1	1	0	0	0	1	0	0	7
1892	1	1	0	1	0	0	0	1	0	0	1	1	0	6
1896	1	1	1	0	1	1	1	1	0	0	1	0	0	8
1912	1	1	0	1	0	0	1	0	0	0	1	1	0	6
1920	1	1	1	0	1	1	0	1	0	1	0	1	0	8
1932	1	0	0	0	1	1	1	1	0	0	1	1	1	8
1952	0	1	1	0	0	1	1	0	1	1	0	1	1	8
1960	1	0	1	0	1	1	1	0	0	1	1	1	1	9
1968	1	1	1	1	0	0	0	1	0	1	1	1	0	8
1976	1	1	0	0	0	1	1	0	1	1	1	1	0	8
1980	1	1	0	1	1	0	1	0	0	1	0	1	1	8

* Electoral vote did not coincide with popular vote results.

we attempt to find the 'recognition rule' that predicts the outcome of a specific election from answers to the same questionnaire.

Algorithm. Given that the learning material consists of only 31 elections, we have chosen the simplest algorithm, 'recognition by the Hamming distance' [9.5]. It may be described briefly as follows.

First, the 'kernel' is determined. This is the set of answers which are encountered more frequently in elections I than in elections C.

To find the kernel the following numbers are calculated:

$$K(i) = \frac{n(i/I)}{n(I)} - \frac{n(i/C)}{n(C)},$$

where $i = 1, 2, \ldots$ is the sequence number of a question from Table 9.1, $n(i/I)$ and $n(i/C)$ show how many elections I and C have the answer 'yes' to question i, and $n(I)$ and $n(C)$ show how many elections I and C are in the learning material.

The answer to question i in the kernel is 'yes' if $K(i) \geq k$ and 'no' if $K(i) \leq -k$, where k is a numerical threshold. If $|K(i)| < k$ the question is not used for recognition.

Second, the distance D between the election and the kernel is calculated as follows:

$$D = \sum_i W(i),$$

where $W(i)$ is the 'weight' of question i:

$$W(i) = \frac{|K(i)|}{\max_i |K(i)|}.$$

Only the answers which are contrary to those in the kernel are included in this summation. If all $W(i)$ are replaced by 1 then D is simply the number of answers which are unfavorable to the incumbent party (the number of 'no' answers).

Third, a rule of recognition is formulated as follows. If $D \leq L$, an election is recognized as I, a win for the incumbent party, and if $D > L$, an election is recognized as C, a win for the challenging party, where L is a certain threshold.

Data Analysis. For purposes of analysis we assumed $W(i) = 1$ and $k = 0.1$. Including all elections from 1860 to 1980 in the questionnaire we obtain a kernel of all zeroes (all answers 'yes'). This follows the intuitive formulation of the questionnaire. The division of elections acording to that kernel is shown in the last column of Table 9.2, which sets the threshold *a posteriori* as $L = 5$.

A central question for pattern recognition is the reliability of the rule of recognition. The rule derived for presidential elections correctly divides past elections into incumbent and challenger victories. Successful forecasts of the 1984, 1988, and 1992 elections provide additional evidence of the rule's reliability. For its statistical evaluation more forecasts are required.

The stability of our results is also established by exploring variations in the questionnaire. We performed a logical exercise similar to the 'seismic history' experiment in [9.1]. Beginning in 1900 we determined the outcome of each successive election using information from all previous elections. Thus we have recreated the position of a forecaster anticipating every election from 1900 through 1980 on the basis of data from the previous elections dating back to 1860. This experiment gives correct results for 19 of 21 elections, indicating that the situations leading to incumbent or challenger victories has changed little during the past 130 years.

A number of additional tests were performed. We found, for example, that results were unchanged when we excluded the five elections in which the number of votes cast for the two main parties differed by 1% or less (1880, 1884, 1888, 1960, 1968) or when we included only twentieth-century elections in the questionnaire.

We also used the algorithm Cora III [9.1] which can formulate composite questions from combinations of the original set. This failed to produce better or additional results. What makes the Hamming distance algorithm preferable is that it is less vulnerable to accidental combinations that have little substantive meaning.

9.3 Midterm Senatorial Elections: Formal Analysis

Analysis of senatorial elections, held between presidential election years, provides a sufficient sample size to obtain statistically verifiable results. The logic and algorithms of the analysis were the same as those used for presidential elections. The work for [9.6] was performed in 1985 and used to forecast the outcomes of the 1986 elections.

The basic data were taken from the history of the three previous midterm elections, those of 1974, 1978, and 1982. Each of the 50 states was included in at least one of these contests. For the analysis, we selected the same set of integral parameters for all states. They are given in Table 9.3.

The answers to the questions from Table 9.3 are given in [9.6]. As before, the wording of the questions implies that 'yes' is preferable for I elections. Accordingly, the kernel again consists of all zeroes.

Figure 9.1 shows the results of applying pattern recognition to the data. These results establish the following decision rule: the incumbent party retains its senatorial seat (I) if $D < 5$ and loses it to the opposition party (C) if $D \geq 5$.

The forecast of the 1986 election published a week in advance provided a test of this decision rule. This forecast is found in Table 9.4, together with the evaluations of predictions by political authorities and the actual results. Our forecast was correct in 30 states of the 34.

We subjected the 1986 results to statistical analysis, considering (a) the correlation between questionnaire results and election outcomes, and (b) the predictive power of the questionnaire.

0	1	2	3	4	5	6	7
			WY82I				
			UT82I	VT82I			
			TX82I	NJ82I			
			TN82I	NE82I			
			RI82I	MO82I			
			PA82I	IN82I			
	WV82I		MN82I	MT78I			
	WI82I		MI82I	KY78I			
	WA82I	OH82I	MA82I	KS78I			
	ND82I	NY82I	HI82I	GA78I			
	MT82I	ME82I	DE82I	DE78I			
	MS82I	MD82I	CT82I	AR78I			
	TN78I	FL82I	AZ82I	AL78I	WY78I		
	OR78I	NM78I	WV78I	PA74I	NM82C		
	WI74I	NC78I	VA78I	OR74I	NJ78C		
	WA74I	ID78I	TX78I	OK74I	NH78C		
SC78I	SC74I	AK78I	RI78I	MD74I	NE78C	CA82I	
IL78I	MO74I	SD74I	LA78I	AR74I	ME78C	UT74I	
LA74I	IN74I	KS74I	NC74I	AK74I	CO78C	OK78C	SD78C
ID74I	IL74I	IA74I	NY74I	MI78C	VT74C	MN78C	MS78C
HI74I	CT74I	CA74I	ND74I	NV82C	KY74C	NV74C	OH74C
GA74I	AL74I	AZ74I	MA78C	IA78C	CO74C	NH74C	FL74C

Fig. 9.1. Division of off-year election returns (1974–1982) by the Hamming distance D from the kernel (i.e., by the number of 'no' answers to the questionnaire in Table 9.3). Each election is represented by the two-letter state abbreviation, the year, and the outcome: I for an incumbent-party victory, C for a challenger-party victory

To test condition (a) we examined the results for all 34 elections. To test condition (b) we excluded elections whose outcome, according to expert opinion, was easily predictable. Specifically, we excluded all elections identified as 'probably secure' in Congressional Review (Table 9.4).

A detailed analysis, presented in [9.6], established the statistical significance of our results both for the 34 elections and the smaller set of less predictable contests.

Our forecast of the 1990 midterm elections, likewise published in advance [9.4], proved correct in 32 of 35 states. Thus 62 of 69 predictions (90%) for the midterm elections of 1986 and 1990 were correct. In the 1990 cycle, moreover, the forecast correctly anticipated the victory of incumbent party candidates in virtually every state, despite widespread press reports that 'anti-incumbent' sentiment was sweeping the nation.

The reliability of the rule of recognition for midterm senatorial elections was also established through the same numerical experiments applied to presidential elections. Detailed results can be found in [9.6]. The sole source of instability lies in the threshold $L = 5$ that divides elections into I and C. Changing the threshold by 1 leads to numerous mistakes, as shown in Fig. 9.1. With only eight parameters this instability is hardly avoidable; it remains when weights $W(i)$ are introduced into the algorithm.

That our prediction was successful despite this instability implies that each parameter is an essential indicator of electoral outcomes. It may also suggest that the parameters are mutually dependent so that a change in one is likely to be accompanied by a change in others. There may yet be tendencies within the electorate that we have not been able to diagnose through additional questions.

We also used the same methods to analyze senatorial elections held in presidential years [9.5]. Although we did not publish forecasts of the 1988 elections ahead of time, we review presidential year elections to make the picture complete.

The first six questions from Table 9.3 remained unchanged for presidential election years. The last two were replaced with the questions 7A and 8A shown in Table 9.3.

The use of this questionnaire with $D = 5$ correctly predicted the results of 32 of 35 senatorial elections in 1988.

Table 9.3. Eight questions for midterm senatorial elections

Senators are elected for a term of six years. Approximately one-third of the senators are re-elected every two years, in some cases in the middle of the presidency, and in others simultaneously with the presidential election.

A 'yes' answer favors the incumbent party. If no more than four answers are 'no' a victory for the incumbent party is predicted. Otherwise a victory for the challenging party is predicted.

Incumbent party candidate:

1   The incumbent party candidate is the sitting Senator.

2   The incumbent party candidate is a major national figure.

3   There was no serious contest for the incumbent party nomination (the candidate collected no less than two-thirds of the vote in the first round).

Incumbent party:

4   The incumbent party won 60% or more of the vote in the previous election.

5   The incumbent party raised at least 10% more money for the campaign than the opposition.

Challenger candidate:

6   The challenger candidate is not a national figure or a past or present Governor or Member of Congress.

7   The challenger candidate is from the same party as the current President.

8   There was no serious contest for the challenger party nomination.

If the senatorial and presidential elections take place in the same year, the last two questions are replaced with:

7A   The questionnaire for the presidential election predicts victory for the incumbent party.

8A   The incumbent party has a majority in the lower house of the state legislature.

Table 9.4. 1986 senatorial election forecast and returns

| State | Answer[1] | | | | | | | | | CR[2] | Fore- | Actual |
	1	2	3	4	5	6	7	8	D		cast[3]	result
HI	0	0	0	0	0	1	0	0	1	PS	*I*	*I*
OH	0	0	0	0	1	0	0	0	1	PS	*I*	*I*
SC	0	0	0	0	0	1	0	0	1	PS	*I*	*I*
UT	0	0	0	0	0	0	1	0	1	PS	*I*	*I*
AK	0	1	0	1	0	0	0	0	2	PS	*I*	*I*
CT	0	1	0	1	0	0	0	0	2	PS	*I*	*I*
KS	0	0	0	0	0	1	1	0	2	PS	*I*	*I*
KY	0	1	0	0	0	1	0	0	2	PS	*I*	*I*
ND	0	1	0	0	0	0	1	0	2	PV	*I*	*C* *
AR	0	1	0	1	0	0	1	0	3	PS	*I*	*I*
CA	0	0	0	1	1	1	0	0	3	V	*I*	*I*
IL	0	1	0	1	0	1	0	0	3	PS	*I*	*I*
IN	0	0	1	1	0	0	1	0	3	PS	*I*	*I*
IA	0	1	0	1	0	0	1	0	3	PS	*I*	*I*
NH	0	1	0	1	0	0	1	0	3	PS	*I*	*I*
OR	0	0	1	1	0	0	1	0	3	PV	*I*	*I*
VT	0	1	0	1	1	0	0	0	3	PV	*I*	*I*
AZ	1	1	0	1	0	0	1	0	4	PS	*I*	*I*
CO	1	1	0	1	1	0	0	0	4	V	*I*	*I*
ID	0	1	0	1	1	0	1	0	4	V	*I*	*I*
LA	1	1	0	0	1	0	0	1	4	HV	*I*	*I*
NY	0	1	0	1	0	1	1	0	4	PS	*I*	*I*
NC	0	1	0	1	1	0	1	0	4	PV	*I*	*C* *
OK	0	1	0	1	1	0	1	0	4	V	*I*	*I*
WA	0	1	0	1	1	0	1	0	4	PV	*I*	*C* *
WI	0	1	0	1	0	1	1	0	4	V	*I*	*I*
AL	0	1	0	1	1	1	1	0	5	V	*C*	*C*
FL	0	1	0	1	1	0	1	1	5	HV	*C*	*C*
GA	0	1	0	1	1	1	1	0	5	PV	*C*	*C*
MO	1	1	0	1	1	0	0	1	5	HV	*C*	*C*
PA	0	1	0	1	1	1	1	0	5	V	*C*	*I* *
MD	1	1	0	0	1	1	1	1	6	HV	*C*	*C*
NV	1	1	0	1	1	0	1	1	6	HV	*C*	*C*
SD	0	1	1	1	1	0	1	1	6	HV	*C*	*C*

[1] Answer 0 = 'yes' and 1 = 'no'.

[2] CR gives Congressional Review assessment of the incumbent party:
  PS probably secure
  PV potentially vulnerable
  V   vulnerable
  HV highly vulnerable

[3] Result *I* is incumbent party victory and *C* is challenger party victory.

* Forecast happened to be wrong.

9.4 Discussion

Self-organization and Predictability. In the natural world, intricate chaotic systems, after appropriate smoothing, often display stable regularities, including predictability. These regularities are difficult, if not impossible, to derive from the behavior of the system's elementary components. Our results for the American political system suggest that American society comprises such a system during presidential and senatorial elections. The hierarchic system of American electoral groups has stable and predictable aggregate-level behavior with a high degree of integration for the entire nation and even for individual states.

This integration occurs despite the contradictory interests and outlooks of electoral groups. The laws governing the outcome of elections have remained stable at the aggregate level from 1860 through 1988, even though three-fourths of today's voters – women, 18 to 20-year-olds, African-Americans, and the great majority of descendants from Latin America, Asia, and Eastern and Southern Europe – were not part of the nineteenth-century electorate. Electoral systems may thus display features similar to large-scale physical systems that likewise exhibit collective behavior comprehensible only at the level of the system as a whole.

What determines collective choice in American presidential elections is the enduring, pragmatic nature of the American electorate. Contrary to the conventional wisdom, our results suggest that issues and ideology, party affiliation, speeches, debates, and advertising count for little or nothing on election day. What matters is the electorate's assessment of how well an incumbent administration has governed the nation. And that assessment is usually clear before the general election campaign even begins. The analysis thus restores the unity between politics and governing that is torn apart in conventional accounts of how elections turn on the strategy and tactics of campaigns.

The parameters developed for presidential elections probe the multiple dimensions of incumbent-party power and performance. The first four parameters primarily gauge conditions that reflect the strength and unity of the party in power. The next seven parameters measure incumbent achievements and failures across a wide range of public concerns. The final two parameters recognize that personality can make a difference in presidential politics, but only when a candidate is either unusually compelling or of heroic stature.

Only one of the thirteen parameters (13) can be influenced directly by the opposition party. This indicates that an incumbent party largely holds its fate in its own hands. Still, many of the parameters may not be within the administrations's control.

Our findings, for example, suggest an explanation of George Bush's 1988 victory that is radically different from the generally accepted version of events. According to the conventional wisdom, after trailing by as many as 17 percentage points in the polls, Bush began a remarkable 'comeback' with his eloquent convention speech (primarily crafted by master speechwriter Peggy Noonan). He then launched a devastating barrage of negative attacks on Mike Dukakis, orchestrated by political adviser Lee Atwater and designed by advertising expert Roger Ailes. When Dukakis failed to respond to charges that he furloughed dangerous criminals and fouled Boston Harbor, Bush surged permanently ahead. Thus, a brilliantly designed – if shallow and vicious – campaign allegedly changed the minds of the voters.

Our conclusions compel a different version of what actually happened in 1988. Based on the record of the previous four years, as measured by the 13 parameters, a Bush victory was apparent long before the public ever heard of speechwriter Peggy Noonan or furloughed rapist Willie Horton.

Six months prior to the election and three months before Bush's alleged comeback, the following forecast was published: "Barring a suddenly stalled economy and a major disaster between now and election day, George Bush is a shoo-in for the presidency, no matter who winds up as the Democratic nominee." [9.3]

The ability to forecast elections prior to campaigns, however, does not mean that candidates can cease campaigning. Campaigns are integral to the political system in which the parameters operate. If one or both parties decided not to take part in the campaign, the nature of the electoral system might change, with consequences that are as yet unclear.

Still, our results do raise the question of what a party should do when historical factors show that it is likely to lose an election. What certainly will not help are the usual attempts to manipulate voters. No master stroke of strategy, advertising trick, or campaign event has ever reversed an unfavorable situation for either the incumbent or the challenging party.

But the disadvantaged party can at least attempt to change the rules of the political game. A party that has the historical odds stacked against it has nothing to lose and perhaps much to gain by running a campaign of candor and substance. For the opposition, it may be the only chance to recover the White House, short of waiting for disaster to befall the president and the country. To run a new kind of campaign, an opposition candidate would have to fire the ad men, cancel the TV spots, and talk straight to the public about how they would govern the country, revealing at least:

- Prospective cabinet appointments,
- Specific legislative changes,
- Concrete taxing and spending plans,
- Drafts of international treaties,
- Alternative solutions to national crises.

If circumstances were against the party in power, the incentive to run an honest and substantive campaign would be reversed. But the results would still be to elevate the shallow and trivial campaigns that have become routine in America's quadrennial contests for president.

As democracy has spread throughout the world, the contagion of a politics based on attack strategies, sound bites, and stage-managed events has followed. A more substantive and honest presidential campaign in the United States might remove this sand from the bearings of democracy worldwide.

The pragmatic electorate also emerges in senatorial elections. Once again, issues, ideology, and campaign events play no role in prediction. But the analysis of senatorial elections also includes no direct performance measures comparable to those included in the presidential analysis. What is most surprising is that neither the national nor the state economy affects the results of senatorial elections.

Voters thus have more diffuse expectations of senators than of presidents. For Senate elections, perceptions of incumbent performance are captured only indirectly through questions about nomination contests, the political stature of opposition-party candidates, and the competition for financial support.

Taken together, our results illustrate the potential to analyze social systems with the same aggregate-level methods used for studying intricate chaotic systems in the natural world. For presidential elections, extremely smoothed parameters describing the social, economic, and political situation in an election year are averaged for the United States as a whole. The same set of parameters is correlated with the outcomes of elections at least since 1860. For senatorial elections, the same set of similarly smoothed parameters are applied to each of the 50 states. At least since 1874, in all states, for each election year, the outcomes of elections for the Senate follow the same basic pattern.

Our conclusions are generally confirmed by early, published forecasts of senatorial elections in 1986 and 1990 and of presidential elections in 1984, 1988, and 1992. The finding that elections can be predicted without reference to campaigns, issues, or ideologies points to the need for a radically revised understanding of how presidential and senatorial elections really work in the United States.

References

9.1 I.M. Gelfand, Sh.A. Guberman, V.I. Keilis-Borok, L. Knopov, F. Press, Ye. Ratsman, I.M. Rotvain, A.M. Sadovski: The conditions of occurrence of strong earthquakes in California and some other regions. Vychyslitelnaya Seismologiya [Computer Seismology] 9, 3–90 (1976) (in Russian)
9.2 K. DeCell: Interview. The Washingtonian, pp. 142–145 (November 1986)
9.3 K. DeCell, A.J. Lichtman: The thirteen keys to the presidency. Madison Books 1990
9.4 K. DeCell, A.J. Lichtman: Interview. The Washingtonian (October 1990)
9.5 A.J. Lichtman, V.I. Keilis-Borok: Pattern recognition applied to presidential elections in the United States 1860–1980: role of integral, social, economic and political traits. Proc. Nat. Acad. Sci. USA 78, 7230–7234 (1981)
9.6 A.J. Lichtman, V.I. Keilis-Borok: Aggregate-level analysis and prediction of midterm senatorial elections in the United States 1974–1986. Proc. Nat. Acad. Sci. USA 86, 1076–1080 (1989)

10. Problems of Predictability in Ethnogenic Studies

L. N. Gumilev and V. Yu. Yermolaev

Predictability must have become a problem to man as soon as he realized the difference between the past, present and future. It is very likely that already the ancestors of modern man from the *Hominidae* family were aware of that difference, although there is still no unambiguous evidence to that effect. The ability to tell the difference between the modalities of time is an established fact for the people of the past three or four thousand years. These wide chronological boundaries are evidence that *Homo sapiens* has rather strong incentives to develop an insight into the future.

We live in a complicated, changing world that imposes quite a few restrictions on our lives. In many cases these make existence neither easier nor more pleasant. However, one should not think that our ancestors had more freedom: the restrictions they had to abide by were different. In a word, the man in the street had no fewer reasons to go to fortune-tellers in the past than he has today to visit celebrities rumored to have extrasensory powers.

It is quite remarkable, though, that declining or growing interest in the future in everyday life is in no way connected with achievements in science and engineering: the twentieth century, an age of high technologies and breathtaking breakthroughs in various fields of knowledge, has at the same time seen the comeback of astrology, palmistry, and the most primitive eschatology. Odd as it may seem, a contemporary college graduate wanting to read his or her future in the stars or in the palm of his or her hand is in no way different from the customers of Nostradamus in the Middle Ages, or the classical Roman asking an augur about the day to come. In all cases we find the same certainty that the future is predictable in principle and that the success of this or that prediction depends entirely on the fortune-teller's skill.

As soon as we leave aside the vulgar understanding of predictability to focus on its scientific interpretation, we must answer a number of fundamental questions. The first one is: What is the subject matter of predictability? Any expert can make forecasts within his or her sphere of competence: the economist studies the situation in the market in a bid to estimate the likely volume of sales and profit, the politician wants to know the line-up of forces in the struggle for power, the doctor seeks to guess the development of a disease, and so on. However, there is something all topics subject to prediction have in common: man tries to foresee how a system will behave in response to external influences. In other words, predictability should in all cases be interpreted as the predictability of an object's behavior. Yet, the understanding of this universality of the category of behavior varies from one field of knowledge to another. The natural scientists say, for good reason, that it is "more natural to discuss *complicated behavior* [italics added – L.G., V.Ye.] rather than complicated systems" [10.8, p. 12].

On the contrary, the humanities (and still more the social sciences) have traditionally given preference to segmentation and classification by modes of production, languages, cultures, and so on, and not by universal mode of conduct. True, segmental descriptions are necessary, but in any case they are not the end, but only a means for understanding the mechanisms controlling the behavior of structurally varied systems. What is deplorable is not the scholars' commitment to the symbols of language or culture, but the fact that many of the attempts to apply the category 'behavior' to the 'featherless biped' are dismissed as heresies. Alas, some of those whose pursuit is science do not share the understanding of science's mission as searching for and explaining universalities in the temporal ocean of diversity. Classical ethnography is no exception to the rule.

At this point we have to deviate from the main topic of this work to clear up a few nuances. Until just recently, domestic ethnography tended to interpret ethnoses as social entites in which the traditional conditions of material production, the boundaries of social–economic formations, and cultural and linguistic determinants play the key role [10.1, pp. 10–15, 246–250; 10.2, pp. 26–31]. The alternative ethnological concept of ethnogenesis with a non-social paradigm is relatively new [10.4, pp. 112–120; 10.5]. The areas of agreement and disagreement of these two approaches to ethnogenesis have already received extensive coverage and seem to require no further comment [10.7, pp. 231–239]. As we discuss ethnoses and ethnogenesis in the following, we will understand them exclusively as ethnological terms.

Ethnogeny defines ethnogenesis as a non-social process in the terrestrial biosphere. The biochemical energy of the biosphere discovered by Vernadsky constitutes the energy basis of ethnogenesis [10.3, p. 284]. In application to ethnogenesis, one of the authors (Gumilev) coined a term to describe the effect of that energy. This term can be rendered in English as 'drive'. Drive brings into being and maintains the existence in the terrestrial biosphere of individual ethnic systems – natural collectives of people having common behavioral stereotypes, and putting themselves in opposition to other such collectives, which proceeds from the feeling of complementarity. This feeling is primary to any ethnos developing the sense of ethnic compatibility [10.5, p. 262]. The nature of complementarity is hypothetically attributed to the fluctuations of the ethnic field [10.5, pp. 291–293].

Like any other system, the ethnic system is hierarchical. The hierarchy of the ethnosphere consists of:

Consortiums – small groups of people who have much in common (e.g., associations of cultural workers, criminal gangs, religious sects);

Sub-ethnoses – subsystems of ethnoses seen as rather big ethnographic or territorial groups with their own customs and cultures and differing from urban people (e.g., old believers, the residents of Russia's northern maritime areas and of Siberia, and intellectuals);

Ethnoses – the main ethnic taxons, which in the social system of ordinates are referred to as peoples, nationalities, or nations;

Super-ethnoses – the highest taxons in the ethnic hierarchy, incorporating a group of ethnoses having a common cultural dominant (e.g., the 'Christian world', the nomadic cultures of Eurasia, the Classical world).

The connection between the biosphere and the ethnic system in the most general terms looks like this. Some factor external to the biosphere (we may suppose it to be cosmic radiation) causes a mutation (drive impulse) in the human population of a certain area. This mutation brings about a hereditary property of drive in the population that is primary to ethnogenesis. Individuals with drive are physiologically normal, but are distinguished by their high ability to absorb the energy of living matter from the environment and to return it in the form of physical work. Since this property can be inherited, the subsequent generations manifest an increase in the 'quantity' of activity, for instance, a greater number of 'events'. These events are quite remarkable. The excess (free) energy of drive is counterpoised to the self-preservation instinct of the individual. People with drive often act to their own detriment and to the detriment of their posterity. However, as they act in coordination with each other and coordinate the activity of all other people, they invest their excess energy into forming and developing a new behavioral structure – a new ethnos [10.5, p. 262].

People, like ethnoses, vary by the degree of their drive, which changes with time. The drive of the individual can correlate with the self-preservation instinct in different ways. If the drive is greater than the instinct, the individual has excess energy and is capable of exerting extraordinary efforts and of actions beyond his or her vital needs for the sake of various abstract ideals: to this class belong the warrior, the novelist, the administrator, the adventurer, the mystic – one can think of many other such types [10.5, pp. 255–259]. If the drive is equal to the instinct, we have before us a balanced, industrious person with no ambition to fight for anything; he or she will prefer to achieve goals gradually, through 'good work' [10.5, pp. 273–274]. Lastly, there is the other extreme: the instinct is greater than the drive. This is the type of person who lacks initiative, loves life, has simple requirements, and is not inclined to work even for the most pragmatic aim or to earn a living. Here belong drug addicts, heavy drinkers, professional beggars, and others [10.5, pp. 275–277]. Any temporary shift in the number of people with different behavioral strategies in the given ethnos determines the general drive level. Generally speaking, the four mentioned terms – ethnos, drive, complementarity, and ethnic drive field – make it possible to describe any process of ethnic behavior at any stage (Fig. 10.1): the upsurge of drive, the acme, the break, the inertial phase, the obscuration, and homeostasis (ethno-terrain balance) [10.5, pp. 355–431].

This inevitably lengthy introduction was necessary to make it quite clear that the ethnological concept of ethnogenesis has an objective basis, essential to forecasting ethnic processes or, to be more exact, human behavior. But is this enough? The very phenomenon of behavior is sufficient to make a prediction. Yet it is necessary to have a clear idea of how its mechanism works and to typologize the system in question. First and foremost, there arises the task of formulating functional distinctions between the ethnos and society. These distinctions are quite clear. Any ethnos is a natural collective that gives the necessary energy to a socially

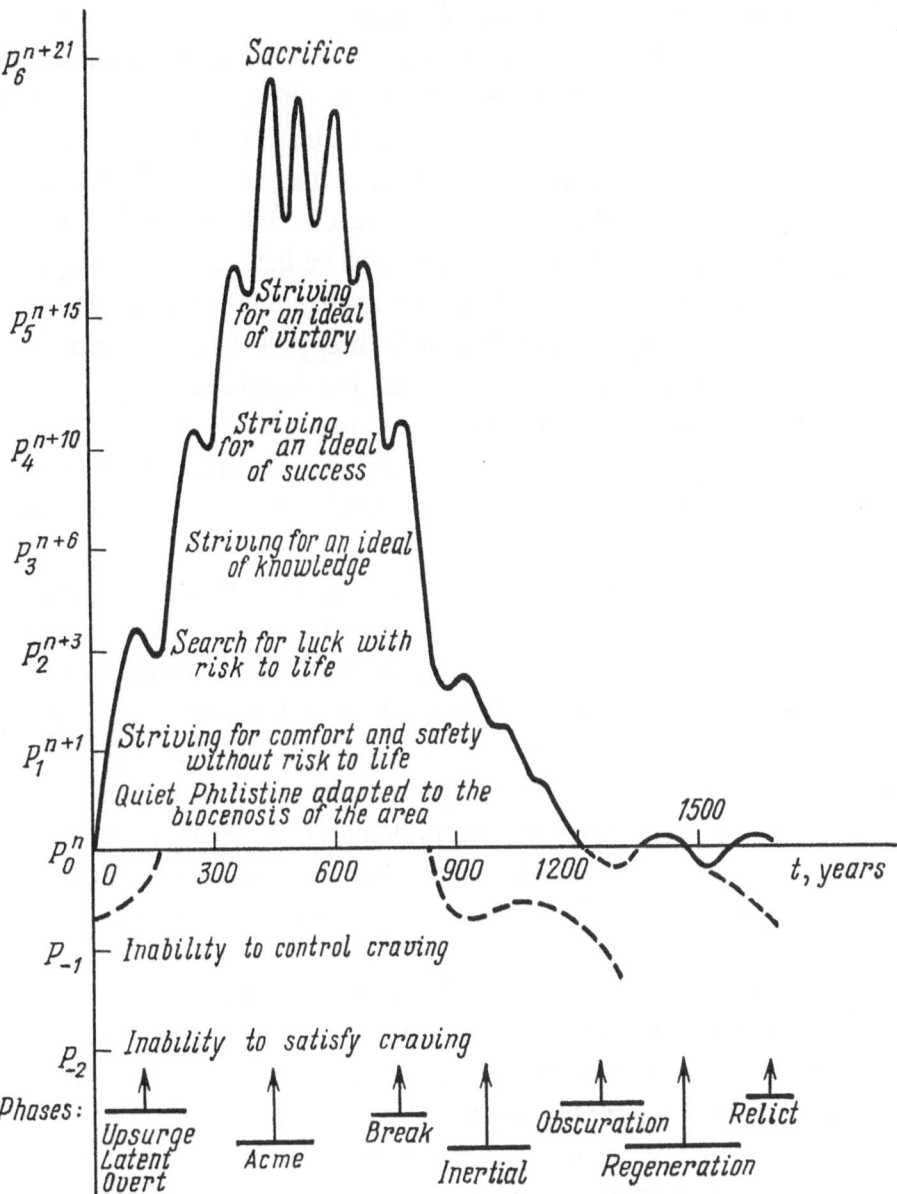

Fig. 10.1. Changes in the drive of an ethnic system with time

strict system that is unable to develop on its own. The interaction of ethnoses and society produces an ethnosocial system, where the ethnos plays an independent and vital role. In other words, it is the ethnos that makes society develop; how this development continues depends directly on the amount of energy the ethnos has lent to society.

Ths history to which we are all witnesses unfolds not within the framework of social–economic formations but within discrete ethnosocial systems. For that reason, the facts of history result from ethnic and social development to an equal degree. Each fact of history reflects social and ethnic dynamics. At the same time, the individual, acting as a student of what happens, is a social creature and can receive information in no way other than through social phenomena. It must be quite clear, though, that the form in which information becomes available is narrower than the whole range of absolute causes that make it appear. In other words, the observed social phenomena have more than just social causes.

Since the ethnosocial system is a marginal, border phenomenon, it should behave accordingly. Can the behavior of the ethnosocial system be likened to the behavior of other marginal structures, using other types of energy?

Let us turn to physics. One of the best-known achievements in this field of knowledge for the past few decades is the laser, a device used for the amplification or generation of coherent light waves with the help of stimulating radiation. Since coherent microwave generation has been discovered in interstellar space, laser action itself can be regarded as lying somewhere in between natural phenomena and man-made technical devices [10.9, p. 21]. This makes the principles of its operation crucial to the topic under discussion. We shall leave aside the mathematical description of these principles to focus on a simpler description, for it is important for us to grasp the general changes in the laser's behavior and to compare them with the behavior of an ethnosocial system.

The general pattern of the generation of a laser impulse is this. An external power source excites the active material of the system, thereby building up its internal energy. The subsequent release of the accumulated energy occurs as a result of fluctuation and produces a visible effect with unique properties, such as high optical power and a high level of coherence of electromagnetic oscillations. Finally, the atoms return to their original, non-excited state. This dynamic behavior of a laser can be paralleled to the behavior of an ethnosocial system.

If we take an ethnosocial system of the super-ethnic rank, the super-ethnos with uniform behavior will constitute its 'active material'. For instance, the ethnosocial system of Western Europe relies on the super-ethnos known as the Christian world (later called the civilized world), whose history can be traced back as far as the ninth century A.D. Any super-ethnic system is non-homogenous and consists of substrates of different origin and different ethnic background. As one can see, the 'active material' here is mixed, too.

Generally speaking, any ethnic system can be in one of two states. The basic (static) state can be described as ethnic homeostasis, where there is no phase dynamics and the stereotype of behavior undergoes no changes. In relation to homeostasis, ethnogenesis is a fluctuation. In the course of ethnogenesis one phase follows another and the stereotype of behavior inherent in the system changes accordingly. Consequently, an ethnic system, just like a laser, has a non-excited state.

The transition from the static state to the dynamic one occurs as a result of a mutagenic influence on the gene pool of the population. One of the authors (Gumilev) maintains that such micro-mutations (drive impulses) can be caused by cosmic radiation [10.5, p. 473], which plays the same role as the external power source in the laser.

Micro-mutation itself can be likened to the charging of the laser medium. The discharge of the medium in the laser is equivalent to the reproduction of the extreme genotype – the property of drive, brought about by the mutation – in the gene pool.

There are some other analogies, too. The mirrors in the laser returning photons to the active medium work just like the system of monogamous restrictions in the super-ethnos. These restrictions limit ethnic contacts and keep up the movement of the entire ethno-social system in a certain direction. This direction is determined by a set of ideas called the ethnic dominant.

The excitement of atoms by an external power source corresponds to the growth of drive in the system: the reproduction of this property and the birth of its carriers – individuals with drive. The build-up of ions at the upper energy levels in the laser's active medium is easily comparable to the acme of ethnogenesis, when the drive of an ethnosocial system is the highest. The quick transition of ions from the excited level to the intermediate level looks very much like the sharp fall in drive during the brief break phase. In relation to the super-ethnos the intermediate level with a long lifetime corresponds to the inertial phase, during which drive changes rather gradually. Lastly, the transition from the intermediate level to the ground state, connected with the main flow of coherent laser radiation, is identical to the fall of drive in the inertial phase of ethnogenesis. This phase results in the emergence of a new socio-cultural entity – 'civilization'.

In either case one can observe a similar dependence of the system's operation on the energy level. When the power source is weak, the laser works as a non-coherent emitter, its radiation is chaotic, and the coherence time is very short. An ethnic system at the very beginning of the upsurge, too, is distinguished by a very low degree of coordination: it exists in the form of consortiums or sub-ethnoses, i.e., taxons whose lifetime is minimal. When the power of the external source is increased to the threshold level, the laser begins to oscillate. Laser radiation appears and the coherence time grows. If the power of the external source is increased still further, the laser's behavior changes again, and it starts acting as a pulsed system [10.9, pp.22–24].

An ethnic system behaves in a very similar way. As soon as drive grows to a certain level, a new ethnos appears together with the appropriate ethosocial system, which coordinates the behavior of the consortiums and sub-ethnoses of the given ethnos. The lifetime of an ethnic entity in this case grows, too, because an ethnos lives much longer than a sub-ethnos and still longer than a consortium.

The subsequent rise in the drive of an ethnic system leads to another development phase. In another acme, drive does change, but it produces nothing like the previous phase of stable growth. There is a quick succession of frequent rises and falls, which is analogous to the generation of short pulses in a laser.

As one can see, the operation of a laser and of an ethnic system are very similar (Table 10.1). From this point of view, ethnogenesis could be called the growth of the living substance of the biosphere caused by exogenous cosmic radiation. This analogy tends to confirm, though indirectly, the hypothesis adopted in ethnology about the extraterrestrial source of mutations [10.5, p. 473].

An example quoted by Haken highlights the typological unity of the behavior of the laser and the ethnosocial system [10.10, Sect. 1.2.2]. This is evidence that the ethnosocial system belongs with the group of dissipative structures that may be described in synergetic terms [10.6, p.31]. However, once the ethnic system is seen as a dissipative structure of the population rank, its behavior is faced with all the restrictions that apply to such structures.

Table 10.1. Analogies in the behavior of a laser and an ethnosocial system

Parameters	Laser	Ethnosocial system
Outside cause	External power source	Hypothetically: cosmic radiation
Target of external influence	Electrons in atoms interacting with the power source and the electro-magnetic field	Gene pool of the population, reproduction of mutagenic properties in posterity
Type of energy	Electromagnetic	Drive – the biochemical energy of the living substance of the biosphere
Active material	Crystal or gas	Ethnic substrates of the previous drive impulses
Growth of active material's potential	Excitement of atoms, followed by population inversion	An increase in the number of individuals with drive, the ethnic system's drive tension grows in the upsurge phase
Active material's maximum potential	Excited particles build up	The acme phase of ethnogenesis
The decline of active material's potential	Transition of particles from the excited level to the intermediate level	A decline in the number of individuals with drive and in the level of drive tension to the optimum
The intermediate level of the decline in the active material's potential	Metastable level	The inertial phase of ethnogenesis
Further decline in the potential of the active material	Transition of particles from the metastable level to the ground (non-excited) level, accompanied by monochrome laser radiation	A fall in the level of drive tension, accompanied by the emergence of an original ethnocultural complex – 'civilization'
Final state of the active material	Ground (non-excited) state	Ethnic balance: homeostasis

In our view, the most serious restriction in studying ethnogenesis is the fundamental impossibility of spotting accurately the points where the ethnic process bifurcates. In ethnogenesis the term bifurcation applies to the moments of phase or sub-phase transition, i.e., changes of trends in the system's drive. The occurrence of such transitions is quite easy to explain. In ethnogenesis there are two basic factors at work. Aging, or the change of a system's drive with time inherent in the reproduction of the gene pool, is one, and the shift which results from the external influence on the given ethnic system of other systems of the same rank is the other. In real history we can never observe pure ethnogenesis. What we can see is the continuous interference of ethnic contacts among representatives of various ethnic systems. Contacts, too, require a rather high degree of drive. As a result, the orthogenetic development of ethnic systems is upset; these systems find new, unprecedented bearings in their development, which are consonant with the new trend in the drive of the system proper.

The mechanism of phase transitions in ethnogenesis resembles the phenomenon of hysteresis. Drive in an ethnic system can have two critical levels. If drive is on the ascent first, the rise may be followed by a fall upon the achievement of the first critical level. This may be a result of mutual extermination: a massacre may claim the lives of those individuals whose drive is the highest (see Fig. 10.1). A certain easing leaves the survivors the opportunity to choose; at the second (lower) critical level they liberate themselves from the pressure of energetic fellow countrymen and rather quickly (in a leap) shift the system to a new state. The people's behavior begins to be dominated not by outdated stereotypes of behavior, but by new ones, which before seemed not only unacceptable but also simply inconceivable. For instance, during the transition from the upsurge phase to the acme phase the imperative of collective responsibility – "Be what you should be!" – is replaced with a purely individualistic one – "Be yourself!". The subsequent phases have their imperatives, too: the break phase – "Everything will be different!"; the inertial phase – "Be like me!"; and the obscuration phase – "Be like us!" [10.5, p. 475]. What makes a prediction so difficult to make is that it is totally impossible to determine which of these imperatives will gain the upper hand, because the role of chance at the moment of choice is rather great. However, as soon as the choice has been made, that is, a new imperative imposed on the system and accepted by an overwhelming majority of its members, the determinism of ethnic development is back again until the next phase transition.

Just as in the case of other dissipative structures, the choice of an ethnic system at the point of bifurcation depends very much on the system's past. The rapid growth of the governing parameter – drive – during the upsurge implies that an equally steep fall of drive during the decline will be highly probable. Thus, the Arabs in the 6th–7th centuries A.D. and the Mongols in the 12th–13th centuries A.D. experienced rapid drive rises only to fail to survive their victories: the colossal empires with polyethnic populations which they founded proved to be ungovernable, because even the growing drive of the conquerors was insufficient for that. Phase transitions ensued to turn the descendants of the Arab ansars and Mongol khans into oppressed minorities.

Ethnic shifts are induced in the same way, when this or that ethnos cuts off its ties with its original terrain. Ancient Rome is an example. All Roman citizens of the 'heroic era' (the upsurge of drive and acme) were both farmers and warriors who fought with the enemies *pro aris et focis* [for holy shrines and homes]. When incessant wars reduced the number of individuals with drive, the army had to be formed of men with sub-drive, who fought for pay and due to strict discipline. This reform by Marius (107 B.C.) only authorized the already existing situation, in which the land was controlled by landowners and cultivated by slaves. The legionnaires who returned home from military expeditions readily gave away their plots of land, for they saw no reason why they should work once they had conquered provinces and made fortunes. This is why the Romans lost contact with their homelands and surrendered their commanding position as an ethnos already in the inertial phase – the era of Augustus. At the end of that era they became extinct, and failed to survive even as a relict in the homeostatic phase.

This example shows that in studying ethnogenesis we are unable to predict the future with a degree of certainty that would be of any practical value. There is no ethnic future, for the actions that shape this future have not yet happened and one cannot say whether they will take place at all. This is only natural, because actions belong in the realm of individual behavior and are determined by a great variety of casual factors. But does this mean that studying ethnogenesis makes no sense? Far from it. Although we cannot predict the immediate future of ethnogenesis due to phase transitions, we can nevertheless do so for longer periods of time.

For instance, the course of ethnogenesis in contemporary Japan tells us that the ethnos of that country will enter its acme by the end of the 21st or the beginning of the 22nd century on the condition that all other factors remain unchanged. This transition will certainly reshape the line-up of economic, social, and political forces in Asia, the Pacific, and elsewhere. However, we cannot be certain that the Japanese ethnos will enter the acme in 2090, and not in 2105. True, in the lifetime of an ethnos, which may last 1200–1500 years, an error of 30 or even 50 years looks insignificant; however, it is quite important for practical recommendations.

So, is it quite clear that in studying ethnogenesis predictability is limited and has a number of specific traits. We cannot make any positive forecasts and say that certain results can be achieved through certain standards of behavior. At best, we can postulate that the observance of certain restrictions in ethnic behavior increases the probability of prolonging the life of an ethnos. Yet, any prediction can be shattered with the beginning of another phase transition, for its chronology and events are largely accidental. For that reason, forecasting in ethnogenic studies is aimed not at predicting the future, but at giving an insight into a global process in order to make adaptation to it more effective.

References

10.1   Yu.V. Bromlei: Ethnos and ethnography. Moscow: Nauka, 1973 (in Russian)
10.2   Yu.V. Bromlei: Contemporary problems of ethnography. Moscow: Nauka, 1981 (in Russian)
10.3   V.I. Vernadsky: The chemical structure of the Earth and its environment. Moscow: Nauka, 1965 (in Russian)
10.4   L.N. Gumilev: On the subject matter of historical geography: terrain and ethnos. III Vestnik LGU, series on geography and geology, 18 (1965) (in Russian)
10.5   L.N. Gumilev: Ethnogenesis and the biosphere of the Earth. Leningrad: Leningrad State University Press (1989) (in Russian)
10.6   V.Yu. Yermolaev: Self-organization in nature and ethnogenesis. Bulletin of the All-Union Geographic Society 121(1) (1990) (in Russian)
10.7   K.P. Ivanov: Views on ethnography, or are there two teachings on ethnoses in Soviet science. Bulletin of the All-Union Geographic Society 117(3) (1987) (in Russian)
10.8   G. Nicolis, I. Prigogine: Exploring complexity. New York: W.H. Freeman and Co., 1989
10.9   H. Haken: Synergetics. Second edition. Berlin: Springer-Verlag, 1982
10.10   H. Haken: Advanced synergetics. Berlin: Springer-Verlag, 1983

Subject Index

Springer Series in Synergetics

Editor: Hermann Haken

Synergetics, an interdisciplinary field of research, is concerned with the cooperation of individual parts of a system that produces macroscopic spatial, temporal or functional structures. It deals with deterministic as well as stochastic processes.

Springer-Verlag
and the Environment

We at Springer-Verlag firmly believe that an international science publisher has a special obligation to the environment, and our corporate policies consistently reflect this conviction.

We also expect our business partners – paper mills, printers, packaging manufacturers, etc. – to commit themselves to using environmentally friendly materials and production processes.

The paper in this book is made from low- or no-chlorine pulp and is acid free, in conformance with international standards for paper permanency.